Anatomy pocket

Authors: Antje Sander, M.D., Stefan Schwarz, M.D.
Editors: Deborah Lorenz-Struve, D.V.M., Carla Maute, M.D.
Translation into English: Lucius Passani, Ph.D.
Cover: Jaqueline Kühne-Hellmessen, Dipl.-Ing. (FH) Publishing
Publisher: Börm Bruckmeier Publishing LLC, www.media4u.com

Important NOTICE – please read!
This book is based on information from sources believed to be reliable, and every effort
has been made to make the book as complete and accurate as possible and to describe
generally accepted practices based on information available as of the printing date, but
its accuracy and completeness cannot be guaranteed. Despite the best efforts of author
and publisher, the book may contain errors, and the reader should use the book only as a
general guide and not as the ultimate source of information about the subject matter.
This book is not intended to reprint all of the information available to the author or
publisher on the subject, but rather to simplify, complement and supplement other
available sources. The reader is encouraged to read all available material and to consult
the package insert and other references to learn as much as possible about the subject.
This book is sold without warranties of any kind, expressed or implied, and the publisher
and author disclaim any liability, loss or damage caused by the content of this book.
If you do not wish to be bound by the foregoing cautions and conditions, you may return
this book to the publisher for a full refund.

Printed in China through Colorcraft Ltd., Hong Kong
ISBN 978-1-59103-219-9

Preface to the First Edition

Anatomy pocket is anatomy reduced to true essentials. This small book is intended as a succinct, clear, and comprehensible way to learn anatomy quickly. This is why we have not included lengthy explanations, but instead have designed the book so that the material is presented in the form of detailed figures and short, concise texts. This will enable you to learn better and to understand and memorize complicated anatomical structures faster.

Many cross-references throughout particular chapters highlight anatomical relationships. For example, if you need to memorize nerve or vessel branching, mnemonics and illustrated charts in the appendix will help you accomplish this. Furthermore, a carefully planned index allows you to look up and review particular topic areas quickly.

Of course, our Anatomy pocket is not intended only for medical students. It is equally well suited for all medical support professionals, from nurses to paramedics.

Students should not be tempted to use this condensed compendium as their sole reference when preparing for examinations. Rather, it should be treated as a useful outline of anatomy.

In designing the Anatomy pocket, we have made every effort to achieve, to the best of our knowledge, a high standard in regard to technical accuracy, graphic quality, and methodology. However, constructive criticism (write to service@media4u.com) is always welcome.

From the authors and the publisher   June 2004

Additional titles in this series:

Börm Bruckmeier Publishing, LLC on the Internet:
www.media4u.com

Contents

6 Contents

8 Contents

10 Contents

12 Contents

1. Introduction

1.1 General Facts

1.1.1 Anatomical Planes and the Terms for Direction and Location

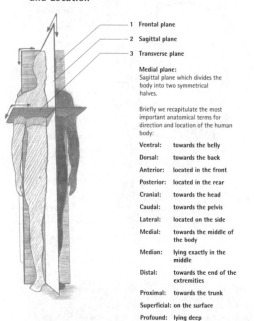

1 Frontal plane

2 Sagittal plane

3 Transverse plane

Medial plane:
Sagittal plane which divides the body into two symmetrical halves.

Briefly we recapitulate the most important anatomical terms for direction and location of the human body:

Ventral:	towards the belly
Dorsal:	towards the back
Anterior:	located in the front
Posterior:	located in the rear
Cranial:	towards the head
Caudal:	towards the pelvis
Lateral:	located on the side
Medial:	towards the middle of the body
Median:	lying exactly in the middle
Distal:	towards the end of the extremities
Proximal:	towards the trunk
Superficial:	on the surface
Profound:	lying deep

2.   Head, Neck

2.1   Basics

2.1.1   Subdivisions of Head and Neck

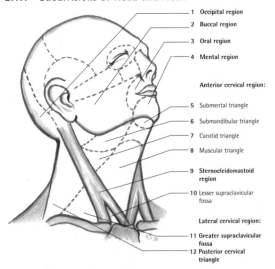

1 Occipital region
2 Buccal region
3 Oral region
4 Mental region

Anterior cervical region:

5 Submental triangle
6 Submandibular triangle
7 Carotid triangle
8 Muscular triangle
9 Sternocleidomastoid region
10 Lesser supraclavicular fossa

Lateral cervical region:

11 Greater supraclavicular fossa
12 Posterior cervical triangle

The numbers 5 to 10 form the **anterior cervical region** and encompass the entire anterior surface of the neck up to the posterior border of the sternocleidomastoid muscle. The submandibular gland is located within the submandibular triangle.

The **lateral cervical region** extends from the posterior border of the sternocleidomastoid muscle to the anterior border of the trapezius muscle.

The **posterior cervical region** is not shown (→ 130).

2.1.2 Fasciae of Neck

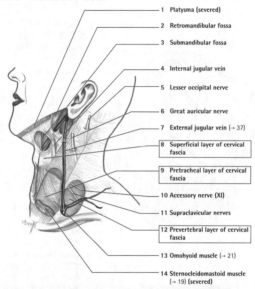

1 Platysma (severed)

2 Retromandibular fossa

3 Submandibular fossa

4 Internal jugular vein

5 Lesser occipital nerve

6 Great auricular nerve

7 External jugular vein (→ 37)

8 Superficial layer of cervical fascia

9 Pretracheal layer of cervical fascia

10 Accessory nerve (XI)

11 Supraclavicular nerves

12 Prevertebral layer of cervical fascia

13 Omohyoid muscle (→ 21)

14 Sternocleidomastoid muscle (→ 19) (severed)

The platysma lies on top of the **superficial layer of cervical fascia** and forms a sheath for the sternocleidomastoid muscle.

The **pretracheal layer of cervical fascia** covers the infrahyoid muscles and extends to the **omohyoid muscle**. It forms a vessel/nerve sheath for the common carotid artery, the internal jugular vein and the vagus nerve.

The infrahyoid muscles can pull on the pretracheal fascia and thereby affect vein lumen width. The **prevertebral layer of cervical fascia** covers the prevertebral neck and scalenus muscles and fuses with the anterior longitudinal ligament of the spine.

2.1.3 Transverse Section of Neck

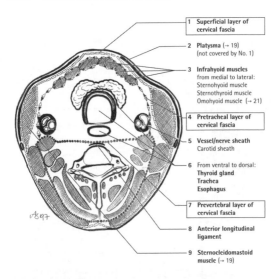

1 **Superficial layer of cervical fascia**

2 **Platysma** (→ 19)
(not covered by No. 1)

3 **Infrahyoid muscles**
from medial to lateral:
Sternohyoid muscle
Sternothyroid muscle
Omohyoid muscle (→ 21)

4 **Pretracheal layer of cervical fascia**

5 **Vessel/nerve sheath**
Carotid sheath

6 From ventral to dorsal:
Thyroid gland
Trachea
Esophagus

7 **Prevertebral layer of cervical fascia**

8 **Anterior longitudinal ligament**

9 **Sternocleidomastoid muscle** (→ 19)

To properly evaluate computer-assisted tomography (CT) images, it is important to identify the cross-sectional structures and to know their spatial relationship to each other!

The platysma is not covered by a fascia. The sternocleidomastoid muscle runs within a sheath formed by the **superficial layer of cervical fascia**. The infrahyoid muscles are covered by the **pretracheal layer of cervical fascia**. This fascial sheet forms the **vessel/nerve sheath**, known as the **carotid sheath**, and encloses the projections of the **common carotid artery, the internal jugular vein and the vagus nerve**.
The **prevertebral layer of cervical fascia** surrounds the prevertebral neck and scalenus muscles (No. 9–11).

2.2 Muscles

2.2.1 Muscles of Neck I

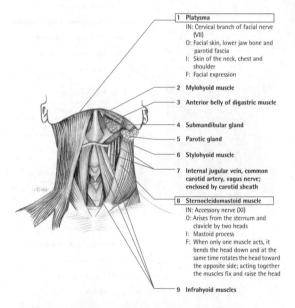

1 **Platysma**
   IN: Cervical branch of facial nerve (VII)
   O: Facial skin, lower jaw bone and parotid fascia
   I: Skin of the neck, chest and shoulder
   F: Facial expression

2 **Mylohyoid muscle**

3 **Anterior belly of digastric muscle**

4 **Submandibular gland**

5 **Parotic gland**

6 **Stylohyoid muscle**

7 **Internal jugular vein, common carotid artery, vagus nerve; enclosed by carotid sheath**

8 **Sternocleidomastoid muscle**
   IN: Accessory nerve (XI)
   O: Arises from the sternum and clavicle by two heads
   I: Mastoid process
   F: When only one muscle acts, it bends the head down and at the same time rotates the head toward the opposite side; acting together the muscles fix and raise the head

9 **Infrahyoid muscles**

When studying muscle function, it is very helpful to deduce the actions of each muscle by looking at its origins and insertions, as well as at the rotational axis of the involved joint.
Infrahyoid muscles: This figure depicts - from medial to lateral - the sternothyroid, sternohyoid and omohyoid muscles.

2.2.2   Muscles of Neck II

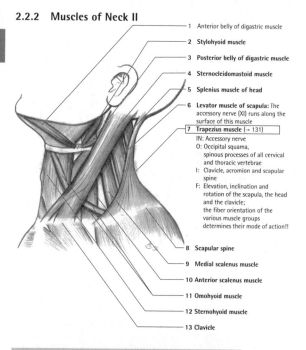

1   Anterior belly of digastric muscle

2   Stylohyoid muscle

3   Posterior belly of digastric muscle

4   Sternocleidomastoid muscle

5   Splenius muscle of head

6   **Levator muscle of scapula:** The accessory nerve (XI) runs along the surface of this muscle

7   **Trapezius muscle (→ 131)**
   IN: Accessory nerve
   O: Occipital squama, spinous processes of all cervical and thoracic vertebrae
   I: Clavicle, acromion and scapular spine
   F: Elevation, inclination and rotation of the scapula, the head and the clavicle; the fiber orientation of the various muscle groups determines their mode of action!!

8   Scapular spine

9   Medial scalene muscle

10   Anterior scalenus muscle

11   Omohyoid muscle

12   Sternohyoid muscle

13   Clavicle

The scalenus gap (→ 36) is shown in the figure as a bright area between the anterior and medial scalenus muscle. It is bounded caudally by the first rib.
The scalenus gap is traversed by the **subclavian artery** and the **primary fasciculus of brachial plexus.**

The posterior scalenus muscle lies below the medial scalenus muscle.

Shown in the figure as numbers 4, 5, 7 and 8 are the **deep muscles of the lateral neck triangle** which are covered by the **prevertebral layer of cervical fascia.**

2.2.3   Infrahyoid Musculature I

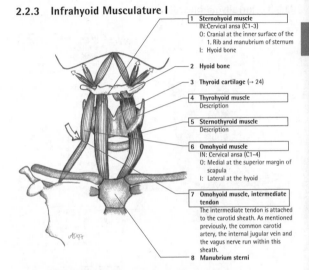

1 Sternohyoid muscle
IN: Cervical ansa (C1-3)
O: Cranial at the inner surface of the
   1. Rib and manubrium of sternum
I: Hyoid bone

2 Hyoid bone

3 Thyroid cartilage (→ 24)

4 Thyrohyoid muscle
Description

5 Sternothyroid muscle
Description

6 Omohyoid muscle
IN: Cervical ansa (C1-4)
O: Medial at the superior margin of
   scapula
I: Lateral at the hyoid

**7 Omohyoid muscle, intermediate
   tendon**
The intermediate tendon is attached
to the carotid sheath. As mentioned
previously, the common carotid
artery, the internal jugular vein and
the vagus nerve run within this
sheath.

8 Manubrium sterni

The sternohyoid, thyrohyoid, sternothyroid and omohyoid muscles together comprise
the **infrahyoid muscles**.
These muscles function to fix and lower the hyoid, to raise the larynx (swallowing act)
and via homohyoid muscle to tense the carotid sheath.
The infrahyoid muscles also partly function as accessory respiratory muscles.

Dissection Information
Caution! Since the inferior belly of the omohyoid muscle is very narrow, it may easily
be confused with a vein.

2.2.4   Infrahyoid Musculature II

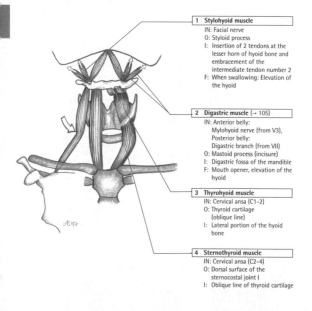

1   Stylohyoid muscle
IN: Facial nerve
O: Styloid process
I: Insertion of 2 tendons at the lesser horn of hyoid bone and embracement of the intermediate tendon number 2
F: When swallowing: Elevation of the hyoid

2   Digastric muscle (→ 105)
IN: Anterior belly: Mylohyoid nerve (from V3), Posterior belly: Digastric branch (from VII)
O: Mastoid process (incisure)
I: Digastric fossa of the mandible
F: Mouth opener, elevation of the hyoid

3   Thyrohyoid muscle
IN: Cervical ansa (C1-2)
O: Thyroid cartilage (oblique line)
I: Lateral portion of the hyoid bone

4   Sternothyroid muscle
IN: Cervical ansa (C2-4)
O: Dorsal surface of the sternocostal joint I
I: Oblique line of thyroid cartilage

The names of the **infrahyoid muscles** (except the omohyoid muscle) can be memorized relatively easy by assuming the hyoid, the thyroid cartilage and the sternum as corner points of a triangle. Each corner point is linked with the other two via one of the muscles. The names of the muscles always refer to their origin and insertion (example: Sternohyoid muscle).

2.2.5 Musculature of Tongue

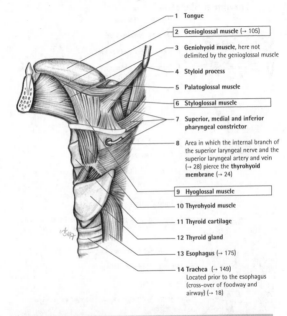

1 Tongue

2 **Genioglossal muscle** (→ 105)

3 **Geniohyoid muscle**, here not delimited by the genioglossal muscle

4 **Styloid process**

5 **Palatoglossal muscle**

6 **Styloglossal muscle**

7 **Superior, medial and inferior pharyngeal constrictor**

8 Area in which the internal branch of the superior laryngeal nerve and the superior laryngeal artery and vein (→ 28) pierce the **thyrohyoid membrane** (→ 24)

9 **Hyoglossal muscle**

10 **Thyrohyoid muscle**

11 **Thyroid cartilage**

12 **Thyroid gland**

13 **Esophagus** (→ 175)

14 **Trachea** (→ 149)
Located prior to the esophagus (cross-over of foodway and airway) (→ 18)

The tongue muscles are divided into **inner** and **outer muscles**.

The **inner tongue muscle group** (→ 105) consists of the **superior and inferior longitudinal muscles**, the **transversus linguae** and the **verticalis linguae**.
The **outer tongue muscle group** is shown in the figure up to the chondroglossus. It comprises the **genioglossus, styloglossus, hyoglossus** and **chondroglossus**.

2.2.6   Ventral and Dorsal View of the Laryngeal Skeleton

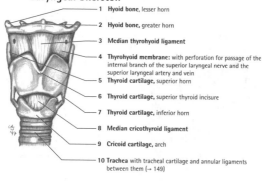

1 **Hyoid bone,** lesser horn

2 **Hyoid bone,** greater horn

3 **Median thyrohyoid ligament**

4 **Thyrohyoid membrane:** with perforation for passage of the internal branch of the superior laryngeal nerve and the superior laryngeal artery and vein

5 **Thyroid cartilage,** superior horn

6 **Thyroid cartilage,** superior thyroid incisure

7 **Thyroid cartilage,** inferior horn

8 **Median cricothyroid ligament**

9 **Cricoid cartilage,** arch

10 **Trachea** with tracheal cartilage and annular ligaments between them (→ 149)

11 **Triticeal cartilage**

12 **Epiglottic cartilage,** the preepiglottic fat body lies ventral to this cartilage; it plays an important role during swallowing

13 **Thyrohyoid membrane** (→ 23)

14 **Lamina of thyroid cartilage**

15 **Corniculate cartilage**

16 **Arytenoid cartilage,** including the vocal and muscular process!

17 **Cricothyroid joint;** enables tilting of the thyroid cartilage on the cricoid cartilage; this process tenses the vocal cords

18 **Lamina of cricoid cartilage**

Below: Once again all the **pharyngeal cartilage:**

Thyroid cartilage, cricoid cartilage, **arytenoid cartilage,** epiglottic cartilage, **corniculate** cartilage, cuneiform cartilage, **triticeal cartilage.**

2.2.7 Muscles of Larynx I

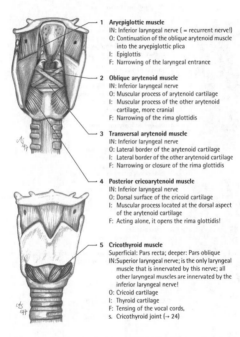

1 **Aryepiglottic muscle**
 IN: Inferior laryngeal nerve (= recurrent nerve!)
 O: Continuation of the oblique arytenoid muscle
  into the aryepiglottic plica
 I: Epiglottis
 F: Narrowing of the laryngeal entrance

2 **Oblique arytenoid muscle**
 IN: Inferior laryngeal nerve
 O: Muscular process of arytenoid cartilage
 I: Muscular process of the other arytenoid
  cartilage, more cranial
 F: Narrowing of the rima glottidis

3 **Transversal arytenoid muscle**
 IN: Inferior laryngeal nerve
 O: Lateral border of the arytenoid cartilage
 I: Lateral border of the other arytenoid cartilage
 F: Narrowing or closure of the rima glottidis

4 **Posterior cricoarytenoid muscle**
 IN: Inferior laryngeal nerve
 O: Dorsal surface of the cricoid cartilage
 I: Muscular process located at the dorsal aspect
  of the arytenoid cartilage
 F: Acting alone, it opens the rima glottidis!

5 **Cricothyroid muscle**
 Superficial: Pars recta; deeper: Pars oblique
 IN: Superior laryngeal nerve; is the only laryngeal
  muscle that is innervated by this nerve; all
  other laryngeal muscles are innervated by the
  inferior laryngeal nerve!
 O: Cricoid cartilage
 I: Thyroid cartilage
 F: Tensing of the vocal cords,
 s. Cricothyroid joint (→ 24)

2.2.8 Muscles of Larynx II

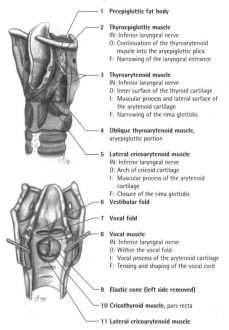

1 **Preepiglottic fat body**

2 **Thyroepiglottic muscle**
   IN: Inferior laryngeal nerve
   O: Continuation of the thyroarytenoid
      muscle into the aryepiglottic plica
   F: Narrowing of the laryngeal entrance

3 **Thyroarytenoid muscle**
   IN: Inferior laryngeal nerve
   O: Inner surface of the thyroid cartilage
   I: Muscular process and lateral surface of
      the arytenoid cartilage
   F: Narrowing of the rima glottidis

4 **Oblique thyroarytenoid muscle,**
   aryepiglottic portion

5 **Lateral cricoarytenoid muscle**
   IN: Inferior laryngeal nerve
   O: Arch of cricoid cartilage
   I: Muscular process of the arytenoid
      cartilage
   F: Closure of the rima glottidis

6 **Vestibular fold**

7 **Vocal fold**

8 **Vocal muscle**
   IN: Inferior laryngeal nerve
   O: Within the vocal fold
   I: Vocal process of the arytenoid cartilage
   F: Tensing and shaping of the vocal cord

9 **Elastic cone (left side removed)**

10 **Cricothyroid muscle,** pars recta

11 **Lateral cricoarytenoid muscle**

Dissection Information

In the figure above the dorsal portion of the thyroid cartilage has been removed.
In the figure below the larynx has been opened posteriorly by cutting through the
cricoid cartilage.

2.2.9   Laryngeal Muscle Function, Frontal Section of Larynx

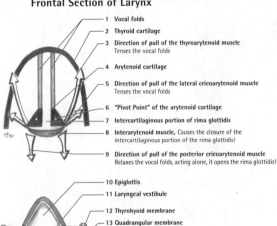

1 Vocal folds

2 Thyroid cartilage

3 Direction of pull of the thyroarytenoid muscle
   Tenses the vocal folds

4 Arytenoid cartilage

5 Direction of pull of the lateral cricoarytenoid muscle
   Tenses the vocal folds

6 "Pivot Point" of the arytenoid cartilage

7 Intercartilaginous portion of rima glottidis

8 Interarytenoid muscle, Causes the closure of the
   intercartilaginous portion of the rima glottidis!

9 Direction of pull of the posterior cricoarytenoid muscle
   Relaxes the vocal folds, acting alone, it opens the rima glottidis!

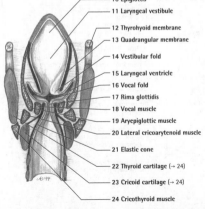

10 Epiglottis

11 Laryngeal vestibule

12 Thyrohyoid membrane

13 Quadrangular membrane

14 Vestibular fold

15 Laryngeal ventricle

16 Vocal fold

17 Rima glottidis

18 Vocal muscle

19 Aryepiglottic muscle

20 Lateral cricoarytenoid muscle

21 Elastic cone

22 Thyroid cartilage (→ 24)

23 Cricoid cartilage (→ 24)

24 Cricothyroid muscle

2.3   Nerves, Vessels

2.3.1   Nerves and Vessels of Larynx

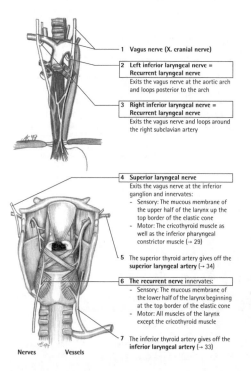

1   **Vagus nerve (X. cranial nerve)**

2   **Left inferior laryngeal nerve = Recurrent laryngeal nerve**
Exits the vagus nerve at the aortic arch and loops posterior to the arch

3   **Right inferior laryngeal nerve = Recurrent laryngeal nerve**
Exits the vagus nerve and loops around the right subclavian artery

4   **Superior laryngeal nerve**
Exits the vagus nerve at the inferior ganglion and innervates:
- Sensory: The mucous membrane of the upper half of the larynx up the top border of the elastic cone
- Motor: The cricothyroid muscle as well as the inferior pharyngeal constrictor muscle (→ 29)

5   The superior thyroid artery gives off the **superior laryngeal artery** (→ 34)

6   **The recurrent nerve innervates:**
- Sensory: The mucous membrane of the lower half of the larynx beginning at the top border of the elastic cone
- Motor: All muscles of the larynx except the cricothyroid muscle

7   The inferior thyroid artery gives off the **inferior laryngeal artery** (→ 33)

Nerves   Vessels

2.4 Swallowing

2.4.1 Constrictor Muscles of the Pharynx

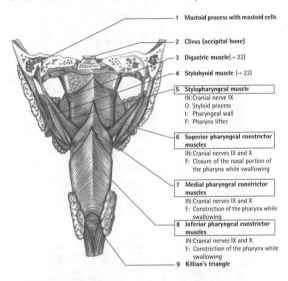

1 Mastoid process with mastoid cells

2 Clivus (occipital bone)

3 Digastric muscle (→ 22)

4 Stylohyoid muscle (→ 22)

5 Stylopharyngeal muscle
IN: Cranial nerve IX
O: Styloid process
I: Pharyngeal wall
F: Pharynx lifter

6 Superior pharyngeal constrictor muscles
IN: Cranial nerves IX and X
F: Closure of the nasal portion of the pharynx while swallowing

7 Medial pharyngeal constrictor muscles
IN: Cranial nerves IX and X
F: Constriction of the pharynx while swallowing

8 Inferior pharyngeal constrictor muscles
IN: Cranial nerves IX and X
F: Constriction of the pharynx while swallowing

9 Killian's triangle

Clinical Information

The pharyngeal constrictor muscles overlap each other and are arranged so that the superior one is innermost and the inferior one is outermost.

The Killian's triangle is the area in which the pharyngeal tube transitions into the esophagus. Due to muscle wall weakness in this area, pulsion diverticula (= formation of a sac or pouch) develop here quite frequently.

It should also be mentioned that the pharyngeal tube is very sensitive to cerebral malfunction (e.g. difficulty in swallowing often develops after apoplexia!)

2.4.2 Swallowing I

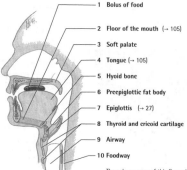

1 Bolus of food

2 Floor of the mouth (→ 105)

3 Soft palate

4 Tongue (→ 105)

5 Hyoid bone

6 Preepiglottic fat body

7 Epiglottis (→ 27)

8 Thyroid and cricoid cartilage

9 Airway

10 Foodway

The sole purpose of this figure is to identify the involved structures. The swallowing process will be illustrated by the following three schematics. The cranial nerves that trigger the individual steps are in brackets.

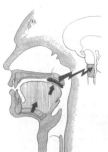

In the **preparatory phase** the tongue pushes **the bolus of food dorsally against the soft palate (XII). The floor of the mouth is raised.** Receptors located in the palate excite the **swallowing center in the medulla oblongata** and swallowing is triggered. The projections of the swallowing center are: Afferent: Cranial nerves V2, IX, X; Efferent: Cranial nerves IX, X, XII. The palatoglossal and palatopharyngeal muscles ensure that the bolus does not slide back from the base of the tongue (IX).

2.4.3 Swallowing II

The next phase involves **automatic actions to keep the food from entering the airways.**
The tensor veli palatini and levator veli palatini muscles block the entrance to the nasal pharynx by lifting and stretching the soft palate (V3, VII, IX). The palatopharyngeal muscle creates a fold in the posterior pharyngeal wall (Passavant's circular ridge).

The **entrance to the larynx** is blocked by two mechanisms:
1. **Upward movement of the hyoid bone** (mylohyoid, stylohyoid, digastric muscles) and of the **thyroid cartilage** toward the hyoid bone (thyrohyoid muscle) moves the larynx entrance closer to the epiglottis (V3, VII, IX);
2. Upward movement of the thyroid cartilage pushes the **fat body** dorsal and moves the epiglottis closer to the larynx entrance.
An automatic temporary **respiratory arrest** is induced.

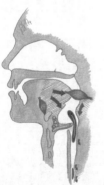

As illustrated in the figure above, the bolus continues its path towards the pharynx.
The **tongue pushes the bolus** via the oropharyngeal isthmus into the pharynx. This process involves the styloglossal and hyoglossal muscles (XII). The pharynx expands.
The **pharyngeal constrictor muscles transport the bolus** through the pharynx (IX). The action of the inferior pharyngeal constrictor muscles causes the formation of a dent in the lower posterior pharyngeal wall. The levator muscles of the pharynx pull the pharyngeal tube over the bolus (IX). The bolus slides **over the epiglottis**, or alternatively, over the epiglottis into the piriform recesses. The pharyngeal constrictor muscles push the food into the **esophagus**, where it is further transported by peristaltic movement.

The swallowing reflex is also present during sleep (vital)

3.   Neck, Chest

3.1   Vessels, Muscles, Nerves

3.1.1   The Subclavian Artery and its Branches

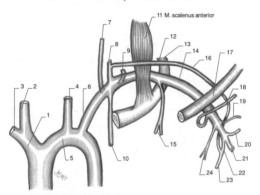

The following structures are depicted:			
1	Brachiocephalic trunk	13	1. rib
2	**Right common carotid artery**	14	Axillary artery (→ 225)
3	**Right subclavian** artery	15	Superior thoracic artery
4	Left common carotid artery	16	Scapular circumflex artery (→ 224)
5	Aortic arch	17	Clavicle
6	**Left subclavian** artery	18	Thoracoacromial artery (→ 225)
7	Vertebral artery (→ 63)	19	Posterior humeral circumflex artery (→ 224)
8	Thyrocervical trunk (→ 33)	20	Anterior humeral circumflex artery (→ 224)
9	Costocervical trunk (→ 33)	21	Brachial artery (→ 224)
10	Internal thoracic artery	22	Subscapular artery (→ 224)
11	Anterior scalenic muscle (→ 35)	23	Thoracodorsal artery (→ 224)
12	Suprascapular artery (→ 224)	24	Lateral thoracic artery

The **subclavian artery** runs towards the 1st rib, at which point it becomes the **axillary artery**. At the lower boundary of the greater pectoral muscle, this vessel becomes the **brachial artery**.
As illustrated in the figure the axillary artery crosses below the clavicle.
The suprascapular artery anastomoses with the scapular circumflex artery ("shoulder blade anastomosis").

3.1.2 Thyrocervical Trunk, Costocervical Trunk

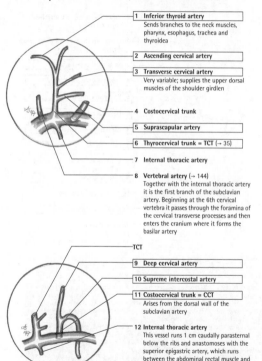

1 Inferior thyroid artery
Sends branches to the neck muscles, pharynx, esophagus, trachea and thyroidea

2 Ascending cervical artery

3 Transverse cervical artery
Very variable; supplies the upper dorsal muscles of the shoulder girdlen

4 Costocervical trunk

5 Suprascapular artery

6 Thyrocervical trunk = TCT (→ 35)

7 Internal thoracic artery

8 Vertebral artery (→ 144)
Together with the internal thoracic artery it is the first branch of the subclavian artery. Beginning at the 6th cervical vertebra it passes through the foramina of the cervical transverse processes and then enters the cranium where it forms the basilar artery

TCT

9 Deep cervical artery

10 Supreme intercostal artery

11 Costocervical trunk = CCT
Arises from the dorsal wall of the subclavian artery

12 Internal thoracic artery
This vessel runs 1 cm caudally parasternal below the ribs and anastomoses with the superior epigastric artery, which runs between the abdominal rectal muscle and the posterior border of the rectus sheath

3.1.3 Branches of External Carotid Artery

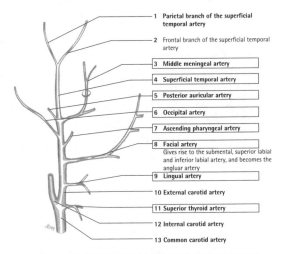

1 Parietal branch of the superficial temporal artery

2 Frontal branch of the superficial temporal artery

3 Middle meningeal artery

4 Superficial temporal artery

5 Posterior auricular artery

6 Occipital artery

7 Ascending pharyngeal artery

8 Facial artery
Gives rise to the submental, superior labial and inferior labial artery, and becomes the angular artery

9 Lingual artery

10 External carotid artery

11 Superior thyroid artery

12 Internal carotid artery

13 Common carotid artery

Clinical Information

The carotid sinus is a small dilated portion of the internal carotid artery above the carotid bifurcation and contains adjacent presso- and chemoreceptors (carotid body).
The pressoreceptors respond to blood pressure, the chemoreceptors to changes in CO_2-, O_2- and H^+-concentration. The afferent impulses of the carotid sinus are transmitted by the glossopharyngeal nerve (IX. cranial nerve).
Carotid sinus syndrome results from hyperactivity of the carotid sinus reflex and leads to bradycardia (eventually leading to cardiac arrest) and hypotension. It is caused by malfunction of the pressoreceptors.
Carotid sinus hyperactivity syndrome is identified by the carotid sinus massage test.
This test measures the response time delay after manual massage of the carotid sinus.

3.1.4 Structures of Deep Lateral Neck Triangle

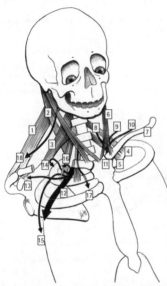

The following muscles serve as guiding structures for locating nerves and vessels within the deep lateral neck triangle: (→ 20)

1 **Trapezius muscle** (→ 20)
2 **Splenius muscle of the neck** (→ 140)
3 **Levator muscle of scapula** (→ 141)
4 **Medial scalenus muscle**
5 **Anterior scalenus muscle**
6 **Sternocleidomastoid muscle**
7 **Clavicle** (partly removed)
8 **Ascending cervical artery** (→ 33)
9 **Transverse cervical artery** (→ 33) Pierces the brachial plexus
10 **Suprascapular artery** (→ 33)
11 **Thyrocervical trunk** (→ 33), arises from the subclavian artery

Important nerves of the lateral neck triangle are:
12 **Brachial plexus** (→ 228)
13 **Suprascapular nerve**
14 **Dorsal nerve of scapula**
15 **Posterior thoracic nerve**
16 **Subclavian nerve**
17 **Phrenic nerve**
18 **Accessory nerve (XI)**
All nerves pass below the clavicle.

The nerves listed above supply the following structures:	
Suprascapular nerve:	**Supraspinatus and infraspinatus muscle** (→ 217)
Dorsal nerve of scapula:	**Levator muscle of scapula, Rhomboid muscles** (→ 132)
Posterior thoracic nerve:	**Serratus anterior** (→ 134)
Subclavian nerve:	**Subclavian muscle**
Phrenic nerve:	**Diaphragm** (→ 155); branches to pericardium, pleura, abdominal area
Accessory nerve:	**Sternocleidomastoid muscle** (→ 19), **Trapezius muscle** (→ 20)

3.1.5 Scalenus Gap and Traversing Structures

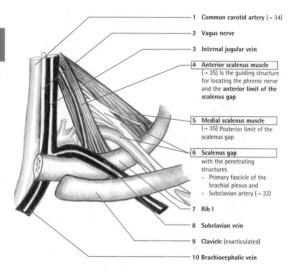

1 **Common carotid artery** (→ 34)

2 **Vagus nerve**

3 **Internal jugular vein**

4 **Anterior scalenus muscle**
(→ 35) Is the guiding structure for locating the phrenic nerve and the **anterior limit of the scalenus gap**

5 **Medial scalenus muscle**
(→ 35) Posterior limit of the scalenus gap

6 **Scalenus gap**
with the penetrating structures
- Primary fascicle of the brachial plexus and
- Subclavian artery (→ 32)

7 **Rib I**

8 **Subclavian vein**

9 **Clavicle** (exarticulated)

10 **Brachiocephalic vein**

The scalenus gap is bounded by the following structures:	
ventral:	Anterior scalenus muscle
dorsocranial:	Medial scalenus muscle
caudal:	1st rib

Also: The subclavian artery, but not the subclavian vein, runs through the scalenus gap! The subclavian vein is located anterior to the scalenus muscle and merges with the internal jugular vein to form the brachiocephalic vein.

3.1.6 Veins of Neck

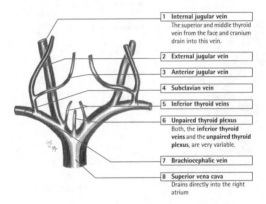

1 Internal jugular vein
The superior and middle thyroid vein from the face and cranium drain into this vein.

2 External jugular vein

3 Anterior jugular vein

4 Subclavian vein

5 Inferior thyroid veins

6 Unpaired thyroid plexus
Both, the **inferior thyroid veins** and the **unpaired thyroid plexus**, are very variable.

7 Brachiocephalic vein

8 Superior vena cava
Drains directly into the right atrium

Dissection Information

Caution! Before dissecting keep in mind that the **external and anterior jugular veins** are located close to the surface!

Clinical Information

Enlarged jugular veins (jugular venous distention) indicate a back-up of venous blood. This may be due to myocardial insufficiency (the influx of venous blood is larger than what the heart can pump out). Alternatively the back-up may be the result of the narrowing of veins emptying into the right atrium of the heart. For example this may be caused by a lung tumor that displaces one of these vessels. Therefore, a quick glance at the neck veins should be part of a patient physical examination!

3.1.7 Thyroid Gland

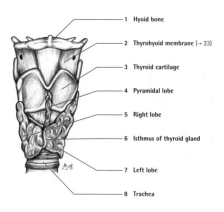

1 Hyoid bone

2 Thyrohyoid membrane (→ 23)

3 Thyroid cartilage

4 Pyramidal lobe

5 Right lobe

6 Isthmus of thyroid gland

7 Left lobe

8 Trachea

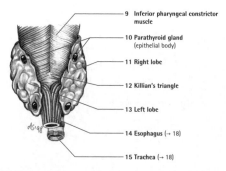

9 Inferior pharyngeal constrictor muscle

10 Parathyroid gland (epithelial body)

11 Right lobe

12 Killian's triangle

13 Left lobe

14 Esophagus (→ 18)

15 Trachea (→ 18)

3.1.8 Vascular Supply of Thyroid

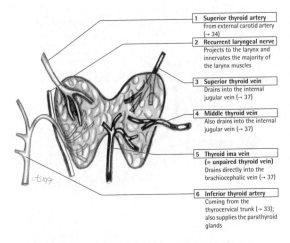

1 Superior thyroid artery
From external carotid artery
(→ 34)

2 Recurrent laryngeal nerve
Projects to the larynx and
innervates the majority of
the larynx muscles

3 Superior thyroid vein
Drains into the internal
jugular vein (→ 37)

4 Middle thyroid vein
Also drains into the internal
jugular vein (→ 37)

5 Thyroid ima vein
(= unpaired thyroid vein)
Drains directly into the
brachiocephalic vein (→ 37)

6 Inferior thyroid artery
Coming from the
thyrocervical trunk (→ 33);
also supplies the parathyroid
glands

Clinical Information

This figure indicates which vessels drain into or arise from a ventral or dorsal position
of the thyroidea.

Thyroid-goiter (knotty enlargement of the thyroidea) surgery is demanding, since
the thyroid gland receives a large number of vessels, which all need to remain
undisturbed. To make matters worse, the recurrent laryngeal nerve runs close by and
must be preserved. Although one-sided recurrent laryngeal paralysis can be survived
(however the patient will be hoarse for the rest of his/her life), two-sided recurrent
laryngeal paralysis leads to failure of the laryngeal vocal cord dilator muscles and to
occlusion of the airways.

3.1.9   Erb's Point

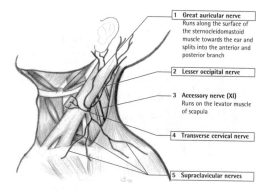

1   **Great auricular nerve**
Runs along the surface of
the sternocleidomastoid
muscle towards the ear and
splits into the anterior and
posterior branch

2   **Lesser occipital nerve**

3   **Accessory nerve (XI)**
Runs on the levator muscle
of scapula

4   **Transverse cervical nerve**

5   **Supraclavicular nerves**

The region at the posterior border of the sternocleidomastoid muscle is named the **Erb's point**
(= **punctum nervosum**). Here the nerves labeled 1, 2, 4 and 5 reach the surface and split in
a star-shaped fashion! These visible nerves are skin nerves arising from the cervical plexus
(→ 42).

Dissection Information

Skin nerves arising from the cervical plexus are usually easy to dissect and easy to prepare.
In some cases the lesser occipital nerve is located below the sternocleidomastoid muscle!
In this case great care should be taken to avoid injury to the accessory nerve!

3.1.10 Important Cervical and Thoracic Structures

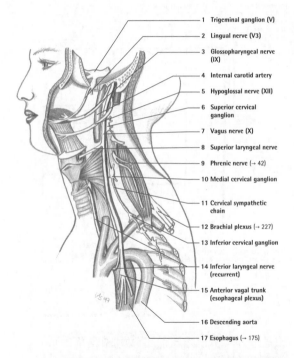

1 Trigeminal ganglion (V)
2 Lingual nerve (V3)
3 Glossopharyngeal nerve (IX)
4 Internal carotid artery
5 Hypoglossal nerve (XII)
6 Superior cervical ganglion
7 Vagus nerve (X)
8 Superior laryngeal nerve
9 Phrenic nerve (→ 42)
10 Medial cervical ganglion
11 Cervical sympathetic chain
12 Brachial plexus (→ 227)
13 Inferior cervical ganglion
14 Inferior laryngeal nerve (recurrent)
15 Anterior vagal trunk (esophageal plexus)
16 Descending aorta
17 Esophagus (→ 175)

3.1.11 Cervical Plexus

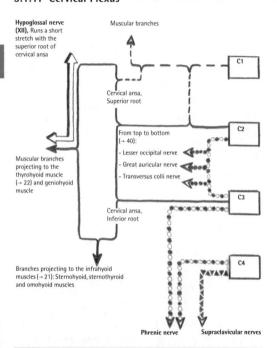

Hypoglossal nerve (XII), Runs a short stretch with the superior root of cervical ansa

Muscular branches

C1

Cervical ansa, Superior root

C2

From top to bottom (→ 40):
- Lesser occipital nerve
- Great auricular nerve
- Transversus colli nerve

Muscular branches projecting to the thyrohyoid muscle (→ 22) and geniohyoid muscle

C3

Cervical ansa, Inferior root

C4

Branches projecting to the infrahyoid muscles (→ 21): Sternohyoid, sternothyroid and omohyoid muscles

Phrenic nerve   **Supraclavicular nerves**

The cervical plexus sends the following branches to the skin: Lesser occipital nerve, great auricular nerve, transverse cervical nerve and supraclavicular nerves (→ 40).

The cervical ansa and the phrenic nerve are branches originating from the cervical plexus.

3.1.12 Sternoclavicular Joint

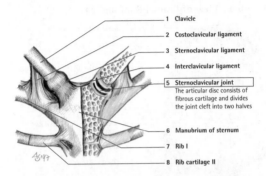

1 Clavicle

2 Costoclavicular ligament

3 Sternoclavicular ligament

4 Interclavicular ligament

5 **Sternoclavicular joint**
The articular disc consists of
fibrous cartilage and divides
the joint cleft into two halves

6 Manubrium of sternum

7 Rib I

8 Rib cartilage II

The **sternoclavicular** joint is a **ball-and-socket joint with three degrees of freedom.**
A degree of freedom is defined as a movement possibility into one specific direction.
At first it seems astonishing that the sternoclavicular joint has three degrees of freedom.
However, this becomes less surprising if one takes a closer look at the degree of
involvement of the clavicle during shoulder movements (you can very easily visualize this
on yourself!).

Dissection Information

Detach the clavicle from the sternoclavicular joint and then proceed with the preparation
of the scalenus gap structures. This procedure is to be performed by the medical assistant.

4.   Cranium, CNS

4.1   Neurocranium

4.1.1   Above View of Cranial Roof and Base

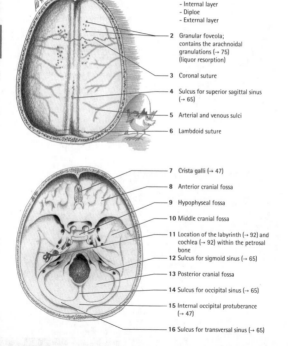

1   **Bones of the cranium with:**
    - Internal layer
    - Diploe
    - External layer

2   Granular foveola;
    contains the arachnoidal
    granulations (→ 75)
    (liquor resorption)

3   Coronal suture

4   Sulcus for superior sagittal sinus
    (→ 65)

5   Arterial and venous sulci

6   Lambdoid suture

7   Crista galli (→ 47)

8   Anterior cranial fossa

9   Hypophyseal fossa

10 Middle cranial fossa

11 Location of the labyrinth (→ 92) and
    cochlea (→ 92) within the petrosal
    bone

12 Sulcus for sigmoid sinus (→ 65)

13 Posterior cranial fossa

14 Sulcus for occipital sinus (→ 65)

15 Internal occipital protuberance
    (→ 47)

16 Sulcus for transversal sinus (→ 65)

4.1.2 Openings of Cranial Base I

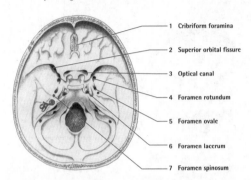

1 Cribriform foramina
2 Superior orbital fissure
3 Optical canal
4 Foramen rotundum
5 Foramen ovale
6 Foramen lacerum
7 Foramen spinosum

	Openings of the cranial base	Penetrating structures	
1	Cribriform foramina	Olfactory nerves (I)	(→ 69)
		A. Anterior ethmoid bone	
2	Superior orbital fissure	Oculomotoric nerve (III)	(→ 69)
		Trochlear nerve (IV)	(→ 70)
		The nasociliary, frontal and lacrimal nerve originate from the ophthalmic nerve (V1)	(→ 70)
			(→ 112)
		Abducent nerve (VI)	(→ 71)
		Orbital branch of the middle meningeal artery	(→ 34)
		Superior ophthalmic vein	(→ 69)
3	Optical canal	Optic nerve (II)	(→ 69)
		Superior ophthalmic artery	
4	Foramen rotundum	Maxillary nerve (V2)	(→ 70)
5	Foramen ovale	Mandibula nerve (V3)	(→ 70)
6	Foramen lacerum	Lesser petrosal nerve originates from the glossopharyngeal nerve (IX)	(→ 72)
7	Foramen spinosum	Middle meningeal artery	(→ 34)
		Meningeal branch of the mandibular nerve (V3)	(→ 70)

4.1.3 Openings of Cranial Base II

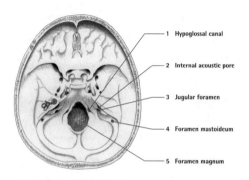

1 Hypoglossal canal

2 Internal acoustic pore

3 Jugular foramen

4 Foramen mastoideum

5 Foramen magnum

	Openings of the cranial base	Penetrating structures	
1	Hypoglossal canal	Hypoglossal nerve (XII)	(→ 73)
2	Internal acoustic pore	Facial nerve (VII)	(→ 71)
		Vestibulocochlear nerve (VIII)	(→ 72)
		Labyrinthine artery with labyrinthine veins	
3	Jugular foramen	Vagus nerve (X)	(→ 72)
		Accessory nerve (XI)	(→ 73)
		Inferior petrosal sinus	
		Sigmoid sinus	(→ 65)
		Posterior meningeal artery originates from the ascending pharyngeal artery	(→ 34)
4	Foramen mastoideum	Mastoid emissary vein	
5	Foramen magnum	Spinal medulla with meninges	(→ 96)
		Spinal roots of the accessory nerve (XI)	(→ 73)
		Vertebral artery	(→ 97)
		Anterior spinal artery	(→ 63)
6	Carotid canal (not shown)	Internal carotid plexus (sympathetic)	(→ 89)
		Internal carotid artery	(→ 63)

4.1.4 Cerebral Falx and Cerebellar Tentorium

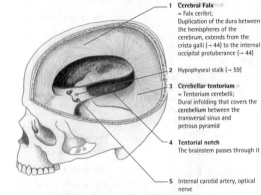

1 **Cerebral Falx**
= Falx ceribri;
Duplication of the dura between the hemispheres of the cerebrum, extends from the crista galli (→ 44) to the internal occipital protuberance (→ 44)

2 **Hypophyseal stalk** (→ 59)

3 **Cerebellar tentorium**
= Tentorium cerebelli;
Dural infolding that covers the cerebellum between the transversal sinus and petrous pyramid

4 **Tentorial notch**
The brainstem passes through it

5 **Internal carotid artery, optical nerve**

The **cranial dura mater** (→ 48) is a protective brain cover. The cerebral falx and the cerebellar tentorium are components that structurally **support parts of the brain**. Also, the fibers of the cerebral falx help to keep the cranial bones in position.

The falx and the tentorium divide the interior of the cranium into a **supratentorial** and **infratentorial compartment**.

In adults the dura attaches primarily to the following structures:	
a) Foramen	d) Lesser wing of sphenoid bone
b) Suture	e) Upper boundary of the petrosal bone
c) Crista galli	f) To the boundaries of the sinuses

Clinical Information

Decerebrate rigidity may develop if a supratentorial space occupying mass within the supratentorial compartment forces portions of the brain into the tentorial notch.
An occupying mass within the infratentorial compartment may cause displacement of portions of the brain through the foramen magnum. This may lead to the compression of the medulla oblongata and to subsequent failure of vital medullary centers (for respiration and circulatory system).

4.2   Cranial Meninges

4.2.1   Layers of the Cranial Meninges

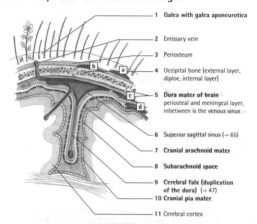

1 Galea with galea aponeurotica

2 Emissary vein

3 Periosteum

4 Occipital bone (external layer, diploe, internal layer)

5 Dura mater of brain periosteal and meningeal layer, inbetween is the venous sinus

6 Superior sagittal sinus (→ 65)

7 Cranial arachnoid mater

8 Subarachnoid space

9 Cerebral falx (duplication of the dura) (→ 47)

10 Cranial pia mater

11 Cerebral cortex

The meningeal and periosteal layer of the dura mater are also referred to as the **pachymeninx**. The two other layers of the dura mater, the arachnoid and the pia mater, are sometimes named the **leptomeninx**. To the meninges of the spinal canal (→ 96).

Clinical Information

Bleeding within the galea and the cranial meninges:	
a	**Subgaleal bleeding** (e.g. trauma during birth)
b	**Subperiosteal bleeding** (cephalohematoma), spread is limited to the boundary of the bones.
c	**Epidural bleeding**; arterial, between the two layers of the dura, strictly localized; often after rupture of the middle meningeal artery.
d	**Subdural bleeding**; venous, between the dura and arachnoid mater, not strictly localized; often after rupture of the bridging veins.
(e)	**Subarachnoid bleeding**; arterial, in the subarachnoid space, often after the bursting of an aneurysm arising from a major intracranial artery.

4.3 Brain

4.3.1 Subdivisions of the Brain I

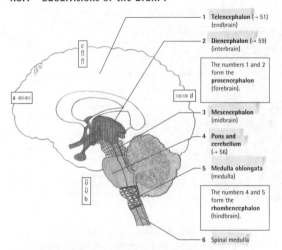

1 **Telencephalon** (→ 51)
(endbrain)

2 **Diencephalon** (→ 59)
(interbrain)

The numbers 1 and 2 form the **prosencephalon** (forebrain).

3 **Mesencephalon** (midbrain)

4 **Pons and cerebellum** (→ 56)

5 **Medulla oblongata** (medulla)

The numbers 4 and 5 form the **rhombencephalon** (hindbrain).

6 Spinal medulla

The brain is divided into the **prosencephalon**, **mesencephalon** and **rhombencephalon**. Combined the pons and cerebellum are called the **metencephalon** (afterbrain). An alternative name for medulla oblongata is **myelencephalon**.

Location of structures within the brain	
a	Oral or rostral
b	Ventral or basal
c	Dorsal
d	Caudal

Each above mentioned brain area is further subdivided into various structures. These are listed briefly on the following page. A more detailed discussion will follow.

4.3.2 Subdivisions of the Brain II

The human brain is divided into the **prosencephalon, mesencephalon** and **rhombencephalon**:			
Prosencephalon	**Diencephalon**	- Thalamus	(→ 60)
		- Hypothalamus	(→ 59)
		- Neurohypophysis	(→ 59)
		- Epiphysis = pineal gland	(→ 62)
		- Globus pallidus	(→ 78)
	Telencephalon	- Cerebral hemispheres	(→ 53)
		- Callosum	(→ 81)
		- Commissural projections	(→ 81)
		- Lateral ventricle	(→ 76)
		- Rhinencephalon (olfactory brain)	
		- Basal ganglia	(→ 78)
Mesencephalon		- Cerebral penduncles	(→ 61)
		- Tectum of mesencephalon (roof of mesencephalon with quadrigeminal plate)	(→ 80)
		- Mesencephalic tegmentum	(→ 80)
		- Mesencephalic aqueduct	(→ 75)
Rhombencephalon	**Metencephalon**	- Pons	(→ 61)
		- Cerebellum with the 3 cerebellar peduncles	(→ 56)
	Myelencephalon	- Medulla oblongata (the transition from the pons to the medulla is gradual)	(→ 61)

The **brain stem** consists of the **mesencephalon, pons** and **medulla oblongata**.

The **basal ganglia**, located in the prosencephalon, are collections of cells that function in combination with other centers to regulate the human motor system.

They include the following nuclei:	
Caudate nucleus	(→ 78)
Putamen (in combination with the caudate nuclei, called the corpus striatum)	(→ 79)
Globus pallidus (in combination with the putamen called the lentiform nuclei)	(→ 78)
In a broader sense the basal ganglia also include the **claustrum** and **amygdaloid nucleus**	(→ 78) (→ 82)

4.3.3 Lobes of the Pallium

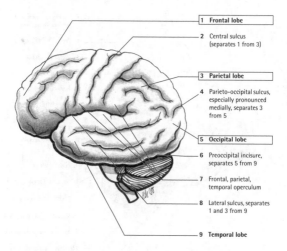

1	**Frontal lobe**
2	Central sulcus (separates 1 from 3)
3	**Parietal lobe**
4	Parieto-occipital sulcus, especially pronounced medially, separates 3 from 5
5	**Occipital lobe**
6	Preoccipital incisure, separates 5 from 9
7	Frontal, parietal, temporal operculum
8	Lateral sulcus, separates 1 and 3 from 9
9	**Temporal lobe**

The **pallium (cerebral cortex)** is subdivided into the lobes listed above. The fifth lobe, the **insula (insular lobe)**, is not visible from the outside. It is located below the lateral sulcus and is covered by portions of the neighboring **opercular lobes**.

There are 3 opercular areas
Frontal operculum
Parietal operculum
Temporal operculum

4.3.4 Important Gyri and Sulci of Pallium

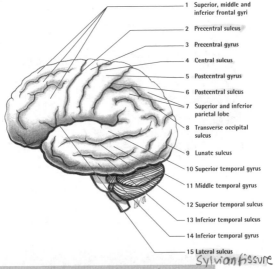

1 Superior, middle and inferior frontal gyri

2 Precentral sulcus

3 Precentral gyrus

4 Central sulcus

5 Postcentral gyrus

6 Postcentral sulcus

7 Superior and inferior parietal lobe

8 Transverse occipital sulcus

9 Lunate sulcus

10 Superior temporal gyrus

11 Middle temporal gyrus

12 Superior temporal sulcus

13 Inferior temporal sulcus

14 Inferior temporal gyrus

15 Lateral sulcus   *Sylvian fissure*

Typical and abundant structures of the **pallium (cerebral cortex)** are the **grooves** ("**Sulci**") and folds ("**Gyri**"). In this figure only the most important gyri and sulci are shown. Briefly we will now introduce functional aspects of the pre- and postcentral gyri. However, a more detailed discussion about location and function of the cerebral cortical areas will follow.

Roughly speaking, the **precentral gyrus** can be seen as the **origin** of the corticospinal tract (pyramidal tract) of the **motor system** (efferent pathway), whereas the **postcentral gyrus** is an important part of the **sensory system** (afferent pathway). However, these two systems are interconnected by various tracts and therefore can not be seen as isolated systems.

4.3.5 Above View of Brain

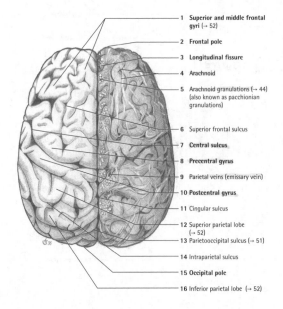

1 **Superior and middle frontal gyri** (→ 52)

2 **Frontal pole**

3 **Longitudinal fissure**

4 **Arachnoid**

5 Arachnoid granulations (→ 44) (also known as pacchionian granulations)

6 Superior frontal sulcus

7 **Central sulcus**

8 **Precentral gyrus**

9 Parietal veins (emissary vein)

10 **Postcentral gyrus**

11 Cingular sulcus

12 Superior parietal lobe (→ 52)

13 Parietooccipital sulcus (→ 51)

14 Intraparietal sulcus

15 **Occipital pole**

16 Inferior parietal lobe (→ 52)

4.3.6 Basal View of Brain

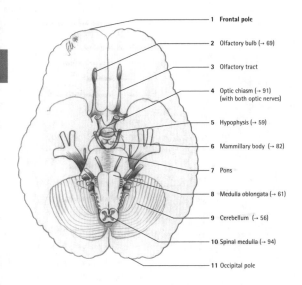

1 Frontal pole

2 Olfactory bulb (→ 69)

3 Olfactory tract

4 Optic chiasm (→ 91)
(with both optic nerves)

5 Hypophysis (→ 59)

6 Mammillary body (→ 82)

7 Pons

8 Medulla oblongata (→ 61)

9 Cerebellum (→ 56)

10 Spinal medulla (→ 94)

11 Occipital pole

4.3.7 Lateral View of Insular Lobe and Medial View of Brain

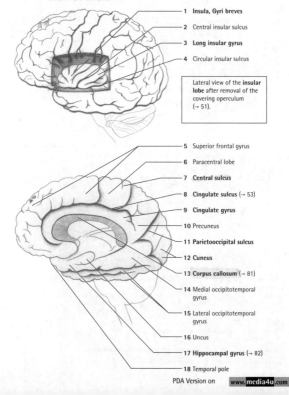

1 Insula, Gyri breves
2 Central insular sulcus
3 **Long insular gyrus**
4 Circular insular sulcus

Lateral view of the **insular lobe** after removal of the covering operculum (→ 51).

5 Superior frontal gyrus
6 Paracentral lobe
7 **Central sulcus**
8 **Cingulate sulcus** (→ 53)
9 **Cingulate gyrus**
10 Precuneus
11 **Parietooccipital sulcus**
12 Cuneus
13 **Corpus callosum** (→ 81)
14 Medial occipitotemporal gyrus
15 Lateral occipitotemporal gyrus
16 Uncus
17 **Hippocampal gyrus** (→ 82)
18 Temporal pole

4.3.8 Cerebellum I

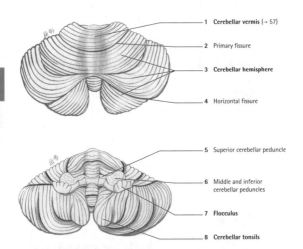

1 Cerebellar vermis (→ 57)

2 Primary fissure

3 Cerebellar hemisphere

4 Horizontal fissure

5 Superior cerebellar peduncle

6 Middle and inferior cerebellar peduncles

7 Flocculus

8 Cerebellar tonsils

A **caudal–rostral view of the cerebellum** is shown in the figure below. The cerebellum embraces the brain stem from its caudal position.

The **function of the cerebellum is to coordinate the motor system**. It compares an internal movement representation plan provided by higher order brain areas with the actual movement (feedback from the movement and vestibular apparatus), and then issues correcting and coordinating efferent signals via the extrapyramidal motor system. The cerebellum affects the temporal sequence of movement, the muscle tone and the balancing of body parts.

The cerebellum is located within the **posterior cranial fossa** and is connected to the brain stem via the **three cerebellar peduncles**. Remember:

The **brainstem** consists of the mesencephalon, pons and medulla oblongata.

4.3.9 Cerebellum II

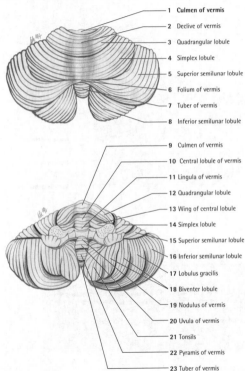

1 **Culmen of vermis**
2 Declive of vermis
3 Quadrangular lobule
4 Simplex lobule
5 Superior semilunar lobule
6 Folium of vermis
7 Tuber of vermis
8 Inferior semilunar lobule

9 Culmen of vermis
10 Central lobule of vermis
11 Lingula of vermis
12 Quadrangular lobule
13 Wing of central lobule
14 Simplex lobule
15 Superior semilunar lobule
16 Inferior semilunar lobule
17 Lobulus gracilis
18 Biventer lobule
19 Nodulus of vermis
20 Uvula of vermis
21 Tonsils
22 Pyramis of vermis
23 Tuber of vermis

4.3.10 Cerebellar Nuclei

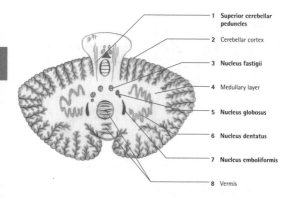

1 Superior cerebellar peduncles
2 Cerebellar cortex
3 **Nucleus fastigii**
4 **Medullary layer**
5 **Nucleus globosus**
6 **Nucleus dentatus**
7 **Nucleus emboliformis**
8 Vermis

As indicated in the figure, the cerebellar cortex consists of many very narrow folds called cerebellar folia, which in transverse cross section appear as pronounced fine arborizations.

Its resemblance to the arborizations of a tree has led to the term **arbor vitae of cerebellum**. The cerebellum consists of the **cerebellar cortex** (= gray matter = nerve cell bodies) and of the central **medullary layer** (= white matter = nerve cell processes), in which the cerebellar nuclei, small collections of nerve cells, are embedded.

The **cerebellar nuclei** include (in bold) the nucleus fastigii, nucleus globosus, nucleus dentatus and the nucleus emboliformis.

4.3.11 Hypothalamus and Pituitary Gland

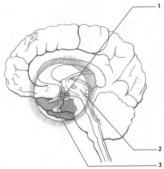

1 Hypothalamus
This lower portion of the Diencephalon **regulates the endocrine glands.** It produces releasing- and inhibiting factors which control the release of glandotropic hormones from the adenohypophysis. It also acts via vegetative nerves directly on the endocrine glands. Furthermore it produces oxytocin and vasopressin in the supraoptic and paraventricular nuclei, which are then both stored in the neurohypophysis. Both, the hypothalamus and the pituitary gland, are connected via a special portal circulatory system. The centers for sleep, hunger and sexuality are all based in the hypothalamus.

2 Thalamus (→ 60)

3 Pituitary gland (→ 61)

4 Pituitary gland
The pituitary gland lies within the hypophyseal fossa, at the base of the cranium (→ 44). It is divided into the ventral **adenohypophysis** and the dorsal **neurohypophysis**. During embryonic development the adenohypophysis develops from the Rathke's pouch of the pharyngeal roof. The adenohypophysis produces (stimulated by hypothalamic releasing factors, that travel via the blood to the adenohypophysis) glandotropic hormones (ACTH, LH, FSH, TH, STH and prolactin).
The neurohypophysis receives the hypothalamic hormones oxytocin and vasopressin via axons and then stores them until they are released into the circulatory system. For surgery, the pituitary gland is accessed through the nasal cavity and the sphenoidal sinus.

4.3.12 Thalamus

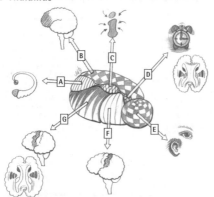

The **thalamus** is located in the diencephalon. It is connected to other parts of the CNS, such as the cerebral cortex, cerebellum, extrapyramidal system and the spinal cord. It is believed to be the **gateway to consciousness**, since it is the area where all sensory stimuli from the environment and internal body are collected and relayed. In addition it is a coordination center. Through its connection to the extrapyramidal motor system it is involved in motor responses, elicited by stimuli it receives. The thalamus is subdivided into various nuclei, each having different connections and functions. In particular these are:

Nucleus	No.	Connection	Function
Anterior	A	To the limbic system via the mammillothalamic tract	Selective attentiveness (suppression of non-relevant sensations)
Medial	B	To the pre-motor region	Influence on personality
Dorsal	C		Integration center
Centro-medial	D	To the caudate nucleus and putamen Pallidum	Arousal, concentration
Posterior	E	To the pulvinar nucleus, medial and lateral geniculate body	Part of the visual- and auditory pathway
Lateral posterior	F	To the postcentral gyrus, spinothalamic tract	
Lateral anterior	G	To the precentral gyrus, basal ganglia	

4.3.13 Lateral (Right) View of Brainstem, Medulla Oblongata

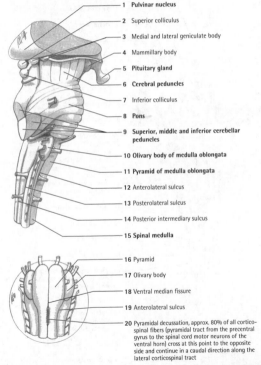

1 Pulvinar nucleus

2 Superior colliculus

3 Medial and lateral geniculate body

4 Mammillary body

5 Pituitary gland

6 Cerebral peduncles

7 Inferior colliculus

8 Pons

9 Superior, middle and inferior cerebellar peduncles

10 Olivary body of medulla oblongata

11 Pyramid of medulla oblongata

12 Anterolateral sulcus

13 Posterolateral sulcus

14 Posterior intermediary sulcus

15 Spinal medulla

16 Pyramid

17 Olivary body

18 Ventral median fissure

19 Anterolateral sulcus

20 Pyramidal decussation, approx. 80% of all cortico-spinal fibers (pyramidal tract from the precentral gyrus to the spinal cord motor neurons of the ventral horn) cross at this point to the opposite side and continue in a caudal direction along the lateral corticospinal tract

4.3.14 Dorsal View of Brainstem

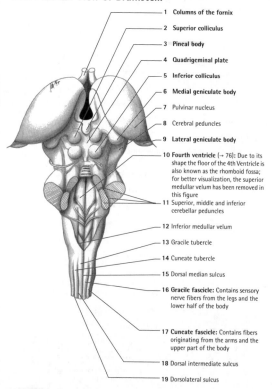

1 Columns of the fornix

2 Superior colliculus

3 Pineal body

4 Quadrigeminal plate

5 Inferior colliculus

6 Medial geniculate body

7 Pulvinar nucleus

8 Cerebral peduncles

9 Lateral geniculate body

10 **Fourth ventricle** (→ 76): Due to its shape the floor of the 4th Ventricle is also known as the rhomboid fossa; for better visualization, the superior medullary velum has been removed in this figure

11 Superior, middle and inferior cerebellar peduncles

12 Inferior medullar velum

13 Gracile tubercle

14 Cuneate tubercle

15 Dorsal median sulcus

16 **Gracile fascicle:** Contains sensory nerve fibers from the legs and the lower half of the body

17 **Cuneate fascicle:** Contains fibers originating from the arms and the upper part of the body

18 Dorsal intermediate sulcus

19 Dorsolateral sulcus

4.4   Intracranial Vessels

4.4.1   Cerebral Arterial Circle

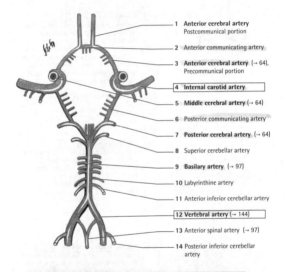

1 **Anterior cerebral artery**
   Postcommunical portion

2 **Anterior communicating artery**

3 **Anterior cerebral artery** (→ 64),
   Precommunical portion

4 **Internal carotid artery**

5 **Middle cerebral artery** (→ 64)

6 **Posterior communicating artery**

7 **Posterior cerebral artery** (→ 64)

8 Superior cerebellar artery

9 **Basilary artery** (→ 97)

10 Labyrinthine artery

11 Anterior inferior cerebellar artery

12 **Vertebral artery** (→ 144)

13 **Anterior spinal artery** (→ 97)

14 Posterior inferior cerebellar
   artery

The **internal carotid artery and the vertebral artery** supply the **cerebral arterial circle** (Willis). Unlike the meningeal vessels, which travel through the epidural cavity (→ 48), **the arteries of the brain run within the subarachnoid space!**

Clinical Information

The communicating arteries are usually not able to compensate, if a large supplying vessel becomes occluded. Thus, cerebral infarction (= ischemic or apoplectic insult) often leads to the demise of brain tissue and subsequently to neurological deficiency. The middle cerebral artery is especially prone to infarction.

4.4.2 Cerebral Arteries

(medial view)

1 Anterior cerebral artery (→ 63)
Supplies the medial surfaces of the cerebral hemispheres up to the parietooccipital sulcus and the dorsal gyri up to approximately 1.5 cm above the superficial margin of the cerebral hemisphere. Furthermore it supplies the rostral portions of the hypothalamus, the caudate nucleus, putamen, globus pallidus, internal capsule and corpus callosum.

2 Parietooccipital sulcus (→ 53)

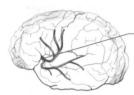

(lateral view)

3 Middle cerebral artery (→ 63)
Supplies the superficial margin as well as the entire outer surface of the cerebral hemispheres, the insular lobe, the internal, external and extreme capsule, the claustrum, the basal ganglia and portions of the thalamus.

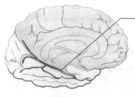

(medial view)

4 Posterior cerebral artery (→ 63)
Supplies the occipital lobe, the basal portion of the temporal lobe, the thalamus, the pineal gland, the globus pallidus, the fornix, the choroid plexus of the 3rd ventricle and the corpus callosum.

4.4.3 Venous Drainage of Brain

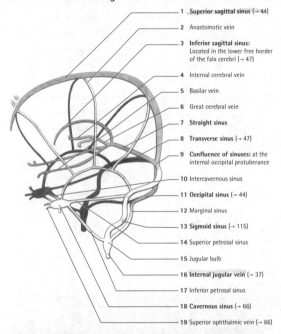

1 **Superior sagittal sinus** (→ 44)

2 Anastomotic vein

3 **Inferior sagittal sinus:**
Located in the lower free border
of the falx cerebri (→ 47)

4 Internal cerebral vein

5 Basilar vein

6 Great cerebral vein

7 **Straight sinus**

8 **Transverse sinus** (→ 47)

9 **Confluence of sinuses:** at the
internal occipital protuberance

10 Intercavernous sinus

11 **Occipital sinus** (→ 44)

12 Marginal sinus

13 **Sigmoid sinus** (→ 115)

14 Superior petrosal sinus

15 Jugular bulb

16 **Internal jugular vein** (→ 37)

17 Inferior petrosal sinus

18 **Cavernous sinus** (→ 66)

19 Superior ophthalmic vein (→ 66)

Sinuses are rigid, venous blood channels. Their lumen is formed by separation of the two layers of the dura mater. They cannot collapse. They receive venous blood from the brain via the so-called bridging veins. The meshwork of the **cavernous sinuses is located laterally to the** sella turcica and is pierced by several **cranial nerves** (→ 66) **and the internal carotid artery.** The sinuses drain via the **internal jugular vein** (→ 115) into the venous system (→ 204).

4.4.4 Cavernous Sinuses

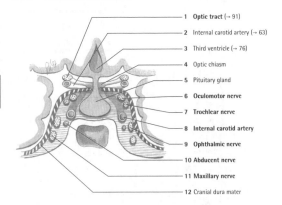

1 **Optic tract** (→ 91)
2 Internal carotid artery (→ 63)
3 Third ventricle (→ 76)
4 Optic chiasm
5 Pituitary gland
6 **Oculomotor nerve**
7 **Trochlear nerve**
8 **Internal carotid artery**
9 **Ophthalmic nerve**
10 **Abducent nerve**
11 **Maxillary nerve**
12 Cranial dura mater

The **cavernous sinuses** are pierced by the **internal carotid artery** and the following cranial nerves:

Oculomotor nerve (III)	(→ 69)
Trochlear nerve (IV)	(→ 70)
Ophthalmic nerve (V1)	(→ 70)
Maxillary nerve (V2)	(→ 70)
Abducent nerve (VI)	(→ 71)

The **abducent nerve** travels within the immediate vicinity of the **internal carotid artery**, while the other above listed cranial nerves are located more laterally within the cavernous sinus. The cavernous sinus receives inflow from the orbital cavity and from the drainage area of the **facial vein** (→ 115) via the superior and inferior ophthalmic veins (→ 65).

Clinical Information

Inflammation occurring above an imaginary line through the corners of the mouth may lead via the ophthalmic veins to intracranial infection!

4.5 Nerves
4.5.1 Cranial Nerves I

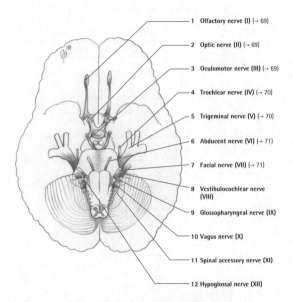

1 Olfactory nerve (I) (→ 69)

2 Optic nerve (II) (→ 69)

3 Oculomotor nerve (III) (→ 69)

4 Trochlear nerve (IV) (→ 70)

5 Trigeminal nerve (V) (→ 70)

6 Abducent nerve (VI) (→ 71)

7 Facial nerve (VII) (→ 71)

8 Vestibulocochlear nerve (VIII)

9 Glossopharyngeal nerve (IX)

10 Vagus nerve (X)

11 Spinal accessory nerve (XI)

12 Hypoglossal nerve (XII)

Mnemonic

On Old Olympus Towering Top A Famous Vocal German Viewed Some Hops.

The bold initial letters of the words in this mnemonic represent the initial letters of the cranial nerves in correct sequence.

4.5.2 Cranial Nerves II (Sites of Emergence from the Brain)

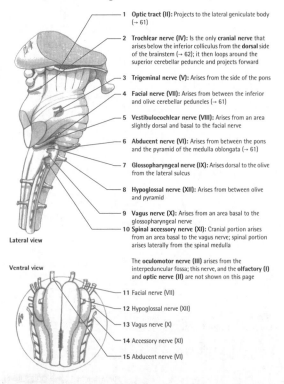

1 **Optic tract (II):** Projects to the lateral geniculate body (→ 61)

2 **Trochlear nerve (IV):** Is the only cranial nerve that arises below the inferior colliculus from the **dorsal** side of the brainstem (→ 62); it then loops around the superior cerebellar peduncle and projects forward

3 **Trigeminal nerve (V):** Arises from the side of the pons

4 **Facial nerve (VII):** Arises from between the inferior and olive cerebellar peduncles (→ 61)

5 **Vestibulocochlear nerve (VIII):** Arises from an area slightly dorsal and basal to the facial nerve

6 **Abducent nerve (VI):** Arises from between the pons and the pyramid of the medulla oblongata (→ 61)

7 **Glossopharyngeal nerve (IX):** Arises dorsal to the olive from the lateral sulcus

8 **Hypoglossal nerve (XII):** Arises from between olive and pyramid

9 **Vagus nerve (X):** Arises from an area basal to the glossopharyngeal nerve

10 **Spinal accessory nerve (XI):** Cranial portion arises from an area basal to the vagus nerve; spinal portion arises laterally from the spinal medulla

Lateral view

Ventral view

The **oculomotor nerve (III)** arises from the interpeduncular fossa; this nerve, and the **olfactory (I)** and **optic nerve (II)** are not shown on this page

11 Facial nerve (VII)

12 Hypoglossal nerve (XII)

13 Vagus nerve (X)

14 Accessory nerve (XI)

15 Abducent nerve (VI)

4.5.3 Cranial Nerves III

Olfactory nerve (I)

General	Contains only **sensory** fibers, thus its fibers project from the periphery to the brain!
Area of perception of stimulus	Olfactory region
Cranial exit	Cribriform plate of ethmoid bone (→ 45)
Course	Olfactory bulb → olfactory tract → medial and lateral olfactory stria → brain area
Brain area	Subcallosal and olfactory area, uncus (limbic system → 82); the sensation of smell strongly affects emotion!)

Optic nerve (II)

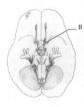

General	Contains only **sensory** fibers, only the nasal fibers of the optic nerve cross at the optic chiasm (the temporal fibers don't!)
Area of perception of stimulus	Retina
Cranial exit	Optic canal (→ 85)
Course	Optic nerve → optic chiasm → optic tract → lateral geniculate body → optic radiation → brain area (→ 91)
Brain area	Visual cortex at the occipital pole (→ 84)

Oculomotor nerve (III)

General	This nerve consists of motor and parasympathetic fibers!
Nuclei	Nucleus of oculomotor nerve (motor), accessory oculomotor nucleus (= Edinger-Westphal nucleus; parasympathetic!) (→ 74)
Cranial exit	Interpeduncular fossa → cranial exit
Course	Superior orbital fissure (→ 85)
Innervated structures	**Motor:** Superior, medial, inferior and inferior oblique muscles of the eye, as well as the levator muscle of the upper eye lid (→ 87) **Parasympathetic:** Ciliary muscle and sphincter muscle of pupil (pupillary constriction)

4.5.4 Cranial Nerves IV

Trochlear nerve (IV)

General	Contains only motor fibers!
Nuclei	Nucleus of the trochlear nerve (motor) (→ 74)
Course	Arises from below the inferior colliculus → wraps around the cerebellar peduncles
Cranial exit	Superior orbital fissure (→ 85)
Innervated structures	Superior oblique muscle (→ 87)

Trigeminal nerve (V)

General	Contains **sensory** and **motor** fibers
Nuclei	Spinal nucleus of trigeminal nerve, pontine nucleus of trigeminal nerve and mesencephalic nucleus of trigeminal nerve (all are sensory) (→ 74) Motor nucleus of trigeminal nerve (motor)
Course	Arises from the side of the pons → trigeminal ganglion → branches into the **ophthalmic nerves (V1), maxillary nerves (V2) and mandibular nerves (V3)** → cranial exits → additional branching occurs (→ 111)
Cranial exit	**V1:** Superior orbital fissure (→ 85) **V2:** Foramen rotundum (→ 45) **V3:** Foramen ovale (→ 45)
Branching	**V1 into:** Frontal nerve, nasociliary nerve, lacrimal nerve (→ 112) **V2 into:** Zygomatic nerve, infraorbital nerve, palatine nerves (→ 112) **V3 into:** Auriculotemporal nerve, inferior alveolar nerve, lingual and masticator nerve (→ 112), innervates the masseter muscles (→ 104)
Area of perception of stimulus / Innervated structure	**Sensory:** Skin of the face, tympanic membrane, oral vestibule **Motor:** Masseter muscles

4.5.5 Cranial Nerves V

Abducent nerve (VI)

General	Contains only **motor** fibers
Nucleus	Nucleus of the abducent nerve (motor) (→ 74)
Course	Arises from between pons and pyramid → cranial exit
Cranial exit	Superior orbital fissure (→ 85)
Innervated structure	Lateral rectus muscle of the eye (→ 87)

Facial nerve (VII)

General	Contains **sensory, motor** and **parasympathetic** fibers!
Nuclei	Solitary and gustatory nucleus (both are sensory), nucleus of the facial nerve (motor) and superior salivary nucleus (parasympathetic) (→ 74)
Course	Arises between olive and inferior cerebellar peduncle → exits through the internal acoustic pore (→ 46) → geniculate ganglion → splits into greater petrosal nerve and facial nerve → facial canal → branches off the chorda tympani nerve → passage through the stylomastoid foramen → innervated structures (→ 103)
Cranial exit	Internal acoustic pore and stylomastoid foramen
Area of perception of stimulus / Innervated structure	**Sensory:** Anterior two-third of the tongue **Motor:** Muscles of facial expression, digastric and stylohyoid muscles, stapedial muscle **Parasympathetic:** lacrimal gland, glands of the nose, sublingual gland

4.5.6 Cranial Nerves VI

Vestibulocochlear nerve (VIII)

General	Consists only of **sensory** fibers!
Area of perception of stimulus	Cochlea and labyrinth (→ 92)
Course	Vestibular branch joins the cochlear branch → passage through the internal acoustic pore → entry between pons and medulla oblongata → **brain nuclei**
Cranial exit	Internal acoustic pore (→ 46)
Nuclei	Ventral and dorsal cochlear nuclei, lateral (Deiter's nucleus), medial (Schwalbe's nucleus), superior (Bechterew's nucleus) and inferior (Roller's nucleus) vestibular nucleus (→ 93)

Glossopharyngeal nerve (IX)

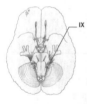

General	Contains **sensory, motor** and **parasympathetic** fibers
Nuclei	Ventral and dorsal cochlear nuclei, lateral (Deiter's nucleus), medial (Schwalbe's nucleus), superior (Bechterew's nucleus) and inferior (Roller's nucleus) vestibular nucleus (→ 74)
Course	Arises dorsal to the olive from the lateral sulcus → cranial exit
Cranial exit	Jugular foramen (→ 46)
Area of perception of stimulus/ Innervated structure	**Sensory:** Back third of the tongue, pharyngeal mucosa **Motor:** Stylopharyngeal muscle and superior pharyngeal constrictor muscle (→ 29) **Parasympathetic:** Parotid gland (→ 103)

4.5.7 Cranial Nerves VII

Vagus nerve (X)

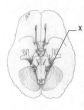

General	Contains **sensory, motor** and **parasympathetic** fibers
Nuclei	Solitary nucleus (sensory), ambiguous nucleus (motor), dorsal nucleus of vagus nerve (parasympathetic) (→ 74)
Course	Arises behind the olive from the lateral sulcus → cranial exit
Cranial exit	Jugular foramen (→ 46)
Area of perception of stimulus / Innervated structure	**Sensory:** Laryngeal mucosa (→ 28) **Motor:** Middle and inferior pharyngeal constrictor muscles (→ 29), levator muscle of palatine velum, laryngeal muscles **Parasympathetic:** Heart, Lung, gastrointestinal tract and others (→ 148)

Spinal accessory nerve (XI)

General	Contains only motor fibers!
Nuclei	Ambiguous nucleus, spinal root nucleus of accessory nerve (motor) (→ 74)
Course	Arises from the side of the medulla oblongata and spinal medulla (→ spinal root ascends through the foramen magnum → joins with the cranial root) → cranial exit → innervated structures
Cranial exit	Jugular foramen (→ 46)
Innervated structure	Sternocleidomastoid muscle (→ 19) and trapezius muscle (→ 131)

Hypoglossal nerve (XII)

General	Contains only motor fibers
Nucleus	Nucleus of hypoglossal nerve (motor) (→ 74)
Course	Arises from between the pyramid and the olive → cranial exit → splits into the lingual branches and into a branch that joins the cervical ansa
Cranial exit	Hypoglossal canal (→ 46)
Innervated structure	Muscles of the tongue (→ 23), infrahyoid muscles (→ 21)

4.5.8 Medial View of Cranial Nerve Nuclei

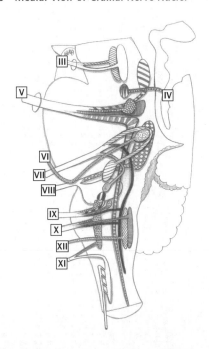

The numbers I–XII represent the 12 cranial nerves. Please refer to the previous pages for more details.

4.6 Subarachnoid Space and Ventricular System

4.6.1 Overview

[handwritten annotations:]
cavities
hydrocephalus: blockage of lateral ventricles by tumor or subarachnoid space block.

1 Subarachnoid space (→ 77)
2 Anterior horn of lateral ventricle, septum pellucidum (thin partition separating the lateral ventricles from each other) (→ 76)
3 Interventricular foramen (of Monro)
[handwritten:] connects lateral ventricles & becomes blocked by tumor
4 Central part of lateral ventricle
5 Interthalamic adhesion
6 Posterior horn of lateral ventricle
7 Temporal horn of lateral ventricle
8 Third ventricle
9 Mesencephalic aqueduct
10 Fourth ventricle
11 Lateral aperture of fourth ventricle (of Luschka)
12 Median aperture of fourth ventricle (of Magendie)
13 Central canal
14 Pontocerebellar cistern

This figure shows the **ventricular system** in dark gray and the **subarachnoid spaces** in light gray. The **cerebrospinal fluid** is produced by the choroid plexus (highly vascularized tissue made from tufts of villi) (→ 76). It flows from the ventricles through the foramina of Luschka and Magendie into the subarachnoidal space and is reabsorbed through the arachnoid granulations (on the roof of the cranium) and the perineurium of brain- and spinal cord nerves. This process replaces the total volume of cerebrospinal fluid contained within the CNS (up to 160 ml) approximately four times a day.

4.6.2 Ventricular System of Brain

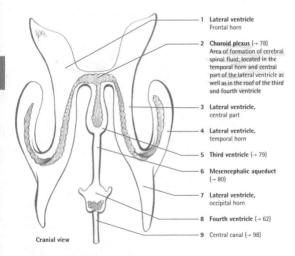

1 **Lateral ventricle**
Frontal horn

2 **Choroid plexus** (→ 78)
Area of formation of cerebral
spinal fluid; located in the
temporal horn and central
part of the lateral ventricle as
well as in the roof of the third
and fourth ventricle

3 **Lateral ventricle,**
central part

4 **Lateral ventricle,**
temporal horn

5 **Third ventricle** (→ 79)

6 **Mesencephalic aqueduct**
(→ 80)

7 **Lateral ventricle,**
occipital horn

8 **Fourth ventricle** (→ 62)

9 **Central canal** (→ 98)

Cranial view

The **ventricular system** of the brain is lined by the **ependyma**, a single layer of
epithelial cells. As is the case for the subarachnoid spaces, it is filled with cerebrospinal
fluid. This clear, water-like liquid consists of 20–40 mg % protein and contains
virtually no cells (max. 5/mm³). It provides a **mechanical cushion** to protect the brain
from impacts and also has **metabolic functions**. By its buoyant action, the
cerebrospinal fluid allows the brain to float, thereby reducing its effective weight on
itself. The cerebrospinal fluid is formed within the **choroid plexus** (approx. 500 ml per
day). By differential hydrostatic pressure it is then driven either via the **arachnoid
granulations** into the blood of the brain sinuses, or alternatively via the **perineurium**
of brain- and spinal cord nerves into the venous system.

4.6.3   Subarachnoid Cisterns

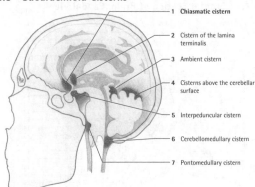

1   **Chiasmatic cistern**

2   Cistern of the lamina terminalis

3   Ambient cistern

4   Cisterns above the cerebellar surface

5   Interpeduncular cistern

6   Cerebellomedullary cistern

7   Pontomedullary cistern

Cranial view of the base of the skull

Cisterns are widening portions of the subarachnoid space.

Here the **basal cistern is seen at the base of the skull in** schematic cranial view.

It consists of the following cisterns: chiasmatic, pontocerebellar, ambient, cerebellomedullary, interpeduncular, lamina terminalis, pontine median, olfactory, carotid, lateral and crural fissure.

Clinical Information

Lumbar puncture is a quite frequently performed clinical procedure for diagnostic and therapeutic purposes. It involves the tapping of the subarachnoid space by insertion of a fine needle, mostly between 3/4 or 4/5 lumbar vertebrae. Alternatively, the cerebellomedullar cistern can also be tapped. This procedure is contraindicated for patients with increased intracranial pressure, since there is the danger of incarceration of brain components within the great foramen, resulting in the compression of vital circulatory and respiratory centers.

4.7   Brain Sections

4.7.1   Frontal Section of Brain, Caudal View

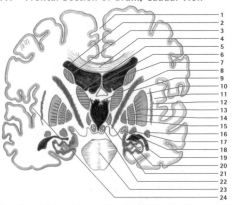

No.	Structure	
1	Cerebral longitudinal fissure	(→ 53)
2	Lateral ventricle	(→ 76)
3	**Caudate nucleus**, head	(→ 79)
4	Corpus callosum, trunk	(→ 81)
5	**Thalamus**	(→ 60)
6	Pellucid septum	(→ 75)
7	Choroid plexus	(→ 76)
8	Choroid plexus	(→ 76)
9	**Internal capsule**	
10	**External capsule**	
11	**Extreme capsule**	
12	**Claustrum**	(→ 79)

Nr.	Struktur	
13	**Putamen**	(→ 79)
14	Lateral division of globus pallidus	
15	Medial division of globus pallidus	
16	Caudate nucleus, tail	
17	**Amygdala**	(→ 82)
18	Optic tract	(→ 91)
19	Lateral ventricle	(→ 76)
20	Hippocampus	(→ 82)
21	**Hypothalamus**	(→ 59)
22	Mammillary body	(→ 61)
23	Pons	(→ 61)
24	Insula	(→ 55)

The **basal ganglia** consist of the caudate nucleus, putamen, pallidum, amygdala and claustrum. The **corpus striatum** is composed of the caudate nucleus and the putamen. Combined the putamen and pallidum form the **lentiform nucleus**. The pallidum comprises the medial and lateral globus pallidus. To properly utilize today's modern imaging technologies (CT, NMR), physicians must be able to identify brain structures in cross-sectional images!

4.7.2 Transverse Section of Brain, View from Above

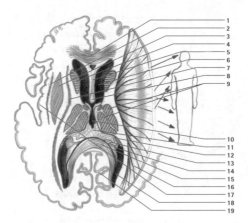

Nr.	Struktur		Nr.	Struktur	
1	**Corpus callosum**, genu	(→ 81)	10	Third ventricle	(→ 75)
2	**Claustrum**	(→ 78)	11	Tela choroidea of 10	
3	**Putamen**	(→ 78)	12	Fornix/Fimbria	(→ 82)
4	**Caudate nucleus**, head	(→ 78)	13	Commissure of fornix	(→ 82)
5	Pellucid septum		14	Hippocampus	(→ 82)
6	Lateral ventricle	(→ 75)	15	**Corpus callosum**, splenium	(→ 81)
7	**Internal capsule**, the somatotopic arrangement is indicated by the arrows	(→ 78)	16	**Internal capsule**, optic radiation	
8	Column of fornix	(→ 82)	17	Insula	(→ 55)
9	**Thalamus**	(→ 60)	18	Lateral ventricle	(→ 75)
			19	Calcarine sulcus	(→ 84)

4.7.3 Transverse Section of Mesencephalon

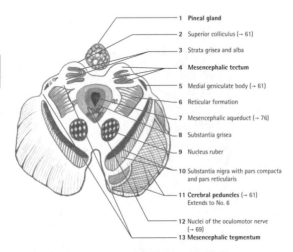

1 Pineal gland

2 Superior colliculus (→ 61)

3 Strata grisea and alba

4 **Mesencephalic tectum**

5 Medial geniculate body (→ 61)

6 Reticular formation

7 Mesencephalic aqueduct (→ 76)

8 Substantia grisea

9 Nucleus ruber

10 Substantia nigra with pars compacta and pars reticularis

11 **Cerebral peduncles** (→ 61) Extends to No. 6

12 Nuclei of the oculomotor nerve (→ 69)

13 **Mesencephalic tegmentum**

The **mesencephalon** comprises the **tectal lamina**, the **tegmentum**, the **tectum** and the **cerebral peduncles**. The ventricular mesencephalic aqueduct traverses the midbrain. The nuclei of the oculomotor and trochlear nerve, as well as the substantia nigra and the nucleus ruber, all lie within the mesencephalon.

Clinical Information

Parkinson's disease, a disorder accompanied by symptoms such as tremors, rigidity and akinesia, is caused by the loss of dopaminergic nerve fibers within the substantia nigra.

4.7.4 Cerebral Fibers, Corpus Callosum

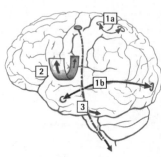

Three major classes of fibers are found within the brain:

1 **Association fibers:**
Connect areas within one hemisphere. They are further divided into the **short arcuate fibers (1a)**, which connect two adjacent gyri with each other, and the **long association fibers (1b)**, which connect 2 lobes within the same hemisphere.

2 **Commissural fibers:**
Interconnect areas of the right and left hemisphere.
There are two types of fibers. The homonymous fibers interconnect two identical centers, whereas the heteronymous fibers link two unlike areas. The most important commissural tract is the corpus callosum **(2)**.

3 **Projection fibers:**
Tracts that originate from within the cortex and project to subcortical nuclei or, as is for example the case for the pyramidal tract, project to the spinal cord **(3)**.

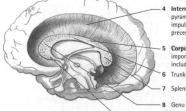

4 **Internal capsule:** Contains the pyramidal tract, which transmits impulses originating from the precentral gyrus.

5 **Corpus callosum**, is the most important commissural tract and includes:

6 Trunk

7 Splenium (→ 79)

8 Genu (→ 79)

9 Rostrum

4.7.5 Limbic System

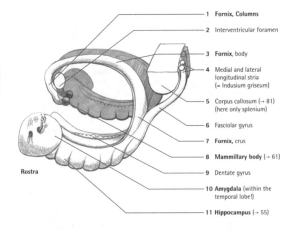

1 **Fornix, Columns**

2 Interventricular foramen

3 **Fornix,** body

4 Medial and lateral
longitudinal stria
(= Indusium griseum)

5 Corpus callosum (→ 81)
(here only splenium)

6 Fasciolar gyrus

7 **Fornix,** crus

8 **Mammillary body** (→ 61)

9 Dentate gyrus

10 **Amygdala** (within the
temporal lobe!)

11 **Hippocampus** (→ 55)

Rostra

The term **limbic system** embraces functionally interrelated cortical areas and nuclei
of the brain. **The limbic system influences, among other things, conscious and
unconscious functions that are important for survival, ingestion, emotion,
sexuality, mental performance and behavior.**

The **cortical components** are divided into an outer and inner ring. The outer ring
consists of the hippocampal and cingulate gyrus (→ 55). The inner ring comprises the
hippocampus, the dentate gyrus, the septal area, the fornix and some additional
smaller structures.

The **non-cortical components** comprise the amygdala, the mammillary body, the
anterior nuclei of the thalamus, the nuclei of the reticular formation and various
others. These components are all members of a **closed circuit** (Papez circuit) with
outgoing and incoming projections. It also should be noted that the olfactory and
limbic system are closely interconnected.

4.7.6 Functional Loops

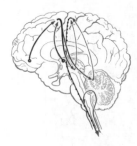

The following three "functional loops" serve as examples to illustrate the complex events that take place within the brain. The figure above depicts the sequence of events that proceed the execution of a movement.

The **motivational movement impulse** arises from the limbic system and the frontal lobe, an area that coordinates motivation and behavior. The next step involves the creation of a movement plan within association areas of the cortex (Brodmann area 6, located prior to the precentral sulcus). The **movement program**, provided by the cerebellum (fast movements) or the basal ganglia (slow movements), is then sent to the precentral gyrus (and also to area 6) via the thalamus. The command to execute the **movement** is subsequently issued from this area.
The ongoing movement is coordinated and fine-tuned by the cerebellum (see previous chapter).

The **function of the basal ganglia** (→ 78) is illustrated by the second example. Their role is to control the sequence, interplay, pace, extent and harmony of complex movements. Impulses from the cerebral cortex are received by the striatum (→ 78) processed by the basal ganglia and then sent out via the globus pallidus and substantia nigra. These modified impulses are relayed back via the thalamus to the cerebral cortex, which then issues the command to execute the movement.

Not shown here is the **cortical processing loop** that lies between **sensation** and **reactive** movement. Here the various stations in sequential order are: Sensory organ - thalamus - primary, secondary and tertiary sensory areas – frontal cortex - tertiary, secondary and primary motor areas – basal ganglia and pyramidal tract - muscle.

4.7.7 Functional Areas of the Cerebral

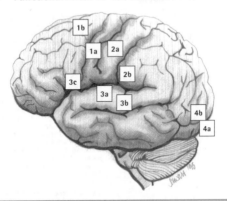

	Area	Localization	
1a	**Motor** projection area	Precentral gyrus	(→ 52)
1b	**Motor** association area	Medial frontal gyrus	(→ 53)
2a	**Sensory** projection area	Postcentral gyrus	(→ 52)
2b	**Sensory** association area	Supramarginal gyrus	
3a	**Auditory** projection area	Superior temporal gyrus	(→ 52)
3a, 3b	**Wernicke's area (language center)** (crucial for language comprehension)	Heschl's gyri	(→ 93)
3c	**Broca's area (language center)** (motor; part of 1a)	Inferior frontal gyrus	(→ 52)
4a	**Visual** projection area at the calcarine sulcus	Occipital pole	(→ 53)
4b	Visual association area at the calcarine sulcus	Occipital pole	(→ 53)

There also exists a classification of the cortical areas according to Brodmann, which distinguishes the different functional areas of the cerebral cortex based on differences in cytoarchitectonics.

4.8   Eye, Orbit

4.8.1   Orbit

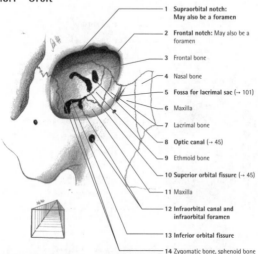

1 **Supraorbital notch:** May also be a foramen

2 **Frontal notch:** May also be a foramen

3 Frontal bone

4 Nasal bone

5 **Fossa for lacrimal sac** (→ 101)

6 Maxilla

7 Lacrimal bone

8 **Optic canal** (→ 45)

9 Ethmoid bone

10 **Superior orbital fissure** (→ 45)

11 Maxilla

12 **Infraorbital canal and infraorbital foramen**

13 **Inferior orbital fissure**

14 **Zygomatic bone, sphenoid bone**

The **eye socket, orbit,** is formed by six cranial bones (→ 44). The four walls, the superior, inferior, lateral and medial wall, form a pyramid whose tip, as indicated above, points in a medial and cranial direction.

Openings of the orbit	Traversing structures
Supraorbital foramen	Branches of the ophthalmic nerve (V1)
Infraorbital foramen	Infraorbital nerve and artery
Superior orbital fissure	Oculomotor nerve (III), trochlear nerve (IV), nasociliary nerve, frontal and lacrimal nerve branching off from the ophthalmic nerve (V1), abducent nerve (VI), orbital branch of the middle meningeal artery, superior ophthalmic vein
Inferior orbital fissure	Zygomatic nerve, artery and vein, infraorbital nerve, artery and vein
Optic canal	**Optic nerv (II),** superior ophthalmic artery

4.8.2 Nerves of Orbit (Frontal View of Frontal Section)

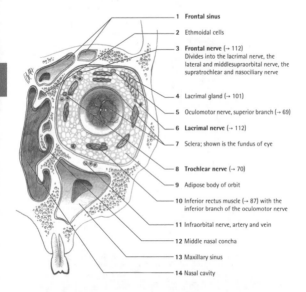

1 **Frontal sinus**

2 Ethmoidal cells

3 **Frontal nerve** (→ 112)
Divides into the lacrimal nerve, the lateral and middlesupraorbital nerve, the supratrochlear and nasociliary nerve

4 Lacrimal gland (→ 101)

5 **Oculomotor nerve, superior branch** (→ 69)

6 **Lacrimal nerve** (→ 112)

7 Sclera; shown is the fundus of eye

8 **Trochlear nerve** (→ 70)

9 Adipose body of orbit

10 Inferior rectus muscle (→ 87) with the inferior branch of the oculomotor nerve

11 Infraorbital nerve, artery and vein

12 Middle nasal concha

13 Maxillary sinus

14 Nasal cavity

The **ophthalmic nerve** (V1) (→ 111) branches into the lacrimal nerve (innervates the lacrimal gland and the skin of the eye corners), the frontal nerve (innervates the skin of the forehead via the supraorbital and supratrochlear nerves) and the nasociliary nerve (which innervates the nasal cavity and the frontal sinus as well as the ethmoidal cells).

The **maxillary nerve** (V2) (→ 111) branches into the infraorbital nerve (sensibility of cheeks and upper jaws) and the zygomatic nerve (sensibility of temporal- and cheekbone area).

4.8.3 Muscles of Eye I

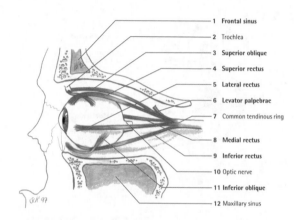

1 Frontal sinus
2 Trochlea
3 Superior oblique
4 Superior rectus
5 Lateral rectus
6 Levator palpebrae
7 Common tendinous ring
8 Medial rectus
9 Inferior rectus
10 Optic nerve
11 Inferior oblique
12 Maxillary sinus

Eye muscle	Innervation
Superior rectus Inferior rectus Medial rectus Inferior oblique	Oculomotor nerve (III)
Superior oblique	Trochlear nerve (IV)
Lateral rectus	Abducent nerve (VI)

Not an eye muscle, but nevertheless important:	
Levator palpebrae superior	Oculomotor nerve (III)

All eye muscles attach at the common tendinous ring!

4.8.4 Muscles of Eye II

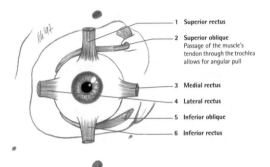

1 **Superior rectus**

2 **Superior oblique**
Passage of the muscle's
tendon through the trochlea
allows for angular pull

3 **Medial rectus**

4 **Lateral rectus**

5 **Inferior oblique**

6 **Inferior rectus**

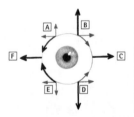

	Eye muscle	Function
A	Inferior oblique	Outward rolling movement, elevation
B	Superior rectus	Elevation
C	Medial rectus	Adduction only
D	Inferior rectus	Depresses
E	Superior oblique	Inward rolling movement, depresses
F	Lateral rectus	Abduction only

4.8.5 Nerves within Orbit, Ciliary Ganglion

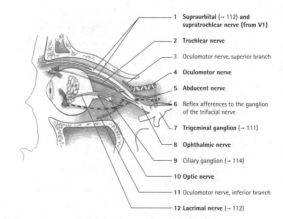

1 **Supraorbital** (→ 112) **and supratrochlear nerve (from V1)**

2 **Trochlear nerve**

3 **Oculomotor nerve, superior branch**

4 **Oculomotor nerve**

5 **Abducent nerve**

6 **Reflex afferences to the ganglion of the trifacial nerve**

7 **Trigeminal ganglion** (→ 111)

8 **Ophthalmic nerve**

9 **Ciliary ganglion** (→ 114)

10 **Optic nerve**

11 **Oculomotor nerve, inferior branch**

12 **Lacrimal nerve** (→ 112)

The **ciliary ganglion** is **parasympathetic** and lies medial to the lateral rectus muscle within the adipose body of orbit. Within this ganglion the preganglionic parasympathetic neurons of the oculomotor nerve are switched to postganglionic ones. These postganglionic fibers combine with sympathetic and sensory fibers and travel as short ciliary nerves to their particular effector organ, e.g. the sphincter muscle of pupil (parasympathetic nervous system causes miosis!) and the ciliary muscle (near-point accommodation with the lens!).

The ciliary ganglion includes the **parasympathetic** and **sympathetic root**, and also fibers derived from the internal carotid plexus (→ 114), which were switched already at the level of the superior cervical ganglion. Therefore these postganglionic fibers pass through the ciliary ganglion without switchover! The third root, the nasociliary root, gives origin to sensory fibers that become part of the nasociliary nerve (→ 112). For these fibers the same applies as for sympathetic fibers: No switchover within the ciliary ganglion!!

4.8.6 Arteries of Orbit, Superior Ophthalmic Artery

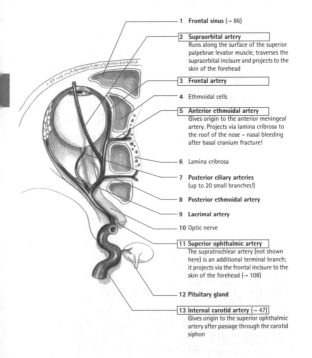

1 **Frontal sinus** (→ 86)

2 **Supraorbital artery**
Runs along the surface of the superior palpebrae levator muscle, traverses the supraorbital incisure and projects to the skin of the forehead

3 **Frontal artery**

4 Ethmoidal cells

5 **Anterior ethmoidal artery**
Gives origin to the anterior meningeal artery. Projects via lamina cribrosa to the roof of the nose – nasal bleeding after basal cranium fracture!

6 Lamina cribrosa

7 **Posterior ciliary arteries**
(up to 20 small branches!)

8 **Posterior ethmoidal artery**

9 **Lacrimal artery**

10 Optic nerve

11 **Superior ophthalmic artery**
The supratrochlear artery (not shown here) is an additional terminal branch; it projects via the frontal incisure to the skin of the forehead (→ 108)

12 Pituitary gland

13 **Internal carotid artery** (→ 47)
Gives origin to the superior ophthalmic artery after passage through the carotid siphon

4.8.7 The Optic Pathway and its Stations

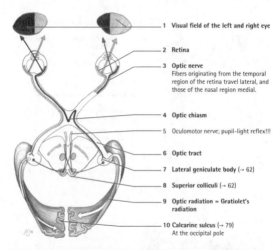

1 **Visual field of the left and right eye**

2 **Retina**

3 **Optic nerve**
Fibers originating from the temporal region of the retina travel lateral, and those of the nasal region medial.

4 **Optic chiasm**

5 **Oculomotor nerve; pupil-light reflex!!!**

6 **Optic tract**

7 **Lateral geniculate body (→ 62)**

8 **Superior colliculi (→ 62)**

9 **Optic radiation = Gratiolet's radiation**

10 **Calcarine sulcus (→ 79)**
At the occipital pole

The stations of the optic pathway:	
1st neuron	Rods and cones in the retina
2nd neuron	Bipolar ganglion cells in the retina
3rd neuron	Cell body in the retina; the axons form the optic nerve
Optic chiasm	Fibers from the nasal half of each retina cross in the chiasm and join uncrossed fibers from the temporal half of the retina to form the optic tract
4th neuron	The cell body is just about always located within the lateral geniculate body; the axons form the optic radiation, which terminates in the visual cortex (calcarine sulcus) (object recognition, movement perception, color vision, spatial perception)
	The 4th neuron may also lie within the superior colliculus (regulation of reflex movements of the eye), the hypothalamus (regulation of the sleep-wake rhythm) or in the pretectal area (regulation of pupil diameter)

4.9 Ear

4.9.1 Ear

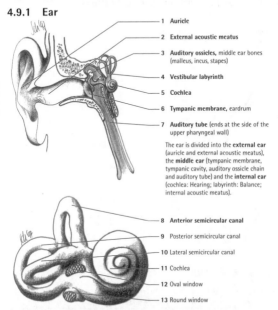

1 **Auricle**

2 **External acoustic meatus**

3 **Auditory ossicles,** middle ear bones (malleus, incus, stapes)

4 **Vestibular labyrinth**

5 **Cochlea**

6 **Tympanic membrane,** eardrum

7 **Auditory tube** (ends at the side of the upper pharyngeal wall)

The ear is divided into the **external ear** (auricle and external acoustic meatus), the **middle ear** (tympanic membrane, tympanic cavity, auditory ossicle chain and auditory tube) and the **internal ear** (cochlea: Hearing; labyrinth: Balance; internal acoustic meatus).

8 **Anterior semicircular canal**

9 Posterior semicircular canal

10 Lateral semicircular canal

11 Cochlea

12 Oval window

13 Round window

An anterolateral view of the right labyrinth is shown in the figure below. The axis of the cochlea within the petrosal bone is orientated in a **posteromedial-superior** to **anterolateral-inferior** direction (→ 44). The sound waves reach the tympanic membrane primarily via the auricle and external acoustic meatus and cause the membrane to vibrate. These vibrations are transmitted via the auditory ossicles to the oval window. The tensor tympani and stapedius muscles can somewhat dampen the transmission of the sound waves (protection of the internal ear from sudden loud sounds). The pressure wave acting on the oval window leads to the displacement of perilymph within the scala vestibuli and scala tympani, and generates a so-called endolymph traveling wave along the endolymphatic duct. The location along the endolymphatic duct, at which this wave has maximum amplitude, is frequency dependent. At this particular location the mechanoelectrical transduction process occurs (cochlear ganglion – cochlear nerve).

4.9.2 Auditory Pathway (Caudal View)

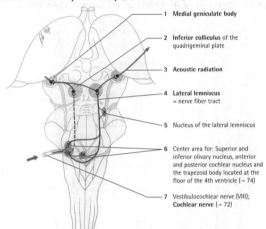

1 Medial geniculate body

2 **Inferior colliculus** of the quadrigeminal plate

3 **Acoustic radiation**

4 **Lateral lemniscus**
 = nerve fiber tract

5 Nucleus of the lateral lemniscus

6 Center area for: Superior and inferior olivary nucleus, anterior and posterior cochlear nucleus and the trapezoid body located at the floor of the 4th ventricle (→ 74)

7 Vestibulocochlear nerve (VIII); **Cochlear nerve** (→ 72)

Stations of the auditory pathway:	
1st neuron	Cell body located in the **cochlear ganglion** (within the cochlea); the axons form the **cochlear nerve.** Fibers from the basal part of the cochlea project to the posterior cochlear nucleus, and those from the apical part project to the anterior cochlear nucleus
2nd neuron	Cell body located in the **cochlear nucleus**; the majority of the axons cross to the opposite side (some establish additional connections within the trapezoid body), to form the **lateral lemniscus**, which projects to the **inferior colliculus**; only very few fibers remain on the ipsilateral side
3rd/4th neuron	Cell body located in the **inferior colliculus**, the axons project primarily to the **medial geniculate body**; also establish a few connections with the cerebellum and superior colliculus
4th/5th neuron	Cell body located primarily in the **medial geniculate body**; the axons form the **acoustic radiation**, whose projections terminate within the **Heschl's gyrus** (a convolution within the temporal lobe, that is oriented towards the insula) or the **Wernicke's language area** of the temporal lobe (→ 84).

4.10 Spinal Cord, Meninges

4.10.1 Dorsal View of Spinal cord (Spinal Medulla)

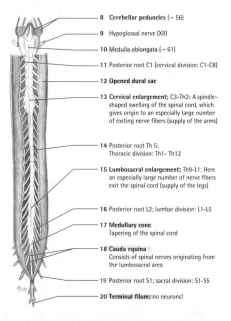

8 **Cerebellar peduncles** (→ 56)

9 **Hypoglossal nerve (XII)**

10 **Medulla oblongata** (→ 61)

11 Posterior root C1 (cervical division: C1-C8)

12 **Opened dural sac**

13 **Cervical enlargement; C3-Th2:** A spindle-shaped swelling of the spinal cord, which gives origin to an especially large number of exiting nerve fibers (supply of the arms)

14 Posterior root Th 5;
Thoracic division: Th1- Th12

15 **Lumbosacral enlargement;** Th9-L1: Here an especially large number of nerve fibers exit the spinal cord (supply of the legs)

16 Posterior root L2; lumbar division: L1-L5

17 **Medullary cone**
Tapering of the spinal cord

18 **Cauda equina**
Consists of spinal nerves originating from the lumbosacral area

19 Posterior root S1; sacral division: S1-S5

20 **Terminal filum;** no neurons!

Shown here is a dorsal view of the **spinal medulla** (spinal cord). The **dural sac has been opened.**

In adults it extends from the atlas to the 1st to 2nd, in children to the 3rd to 4th lumbar vertebrae. Beyond this point only the cauda equina remains.

4.10.2 Nerves of Spinal Cord

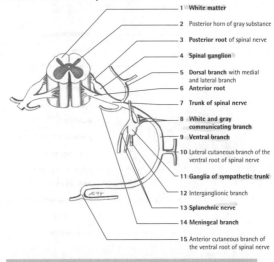

1 **White matter**

2 Posterior horn of gray substance

3 **Posterior root** of spinal nerve

4 **Spinal ganglion**

5 **Dorsal branch** with medial and lateral branch

6 **Anterior root**

7 **Trunk of spinal nerve**

8 **White and gray communicating branch**

9 **Ventral branch**

10 Lateral cutaneous branch of the ventral root of spinal nerve

11 **Ganglia of sympathetic trunk**

12 Interganglionic branch

13 **Splanchnic nerve**

14 **Meningeal branch**

15 Anterior cutaneous branch of the ventral root of spinal nerve

The **posterior and anterior roots** merge at the intervertebral foramen to form the **spinal nerve**. The distance traveled jointly before they divide into the dorsal and ventral branch is called the **trunk of spinal nerve** (approx. 1 cm long). **The posterior (sensory root)** and **anterior root (motor root)** is attached to the spinal cord via the **radicular fila** (very small fibers). The **spinal ganglion** (sensory) lies within the dorsal root. All **31 pairs of spinal nerves** are built alike! This in consequence produces the **segmental innervation of the skin**.

Nerve branch	Innervated Structures
Meningeal branch	Meninges of the spinal cord
Dorsal branch	Skin near the vertebral column, autochthonic muscles of the back
Ventral branch	Remaining skin, muscles of the trunk, neck and extremities

Keep in mind: Only the ventral branch is involved in the formation of each particular plexus!

4.10.3 Meninges of Spinal Canal

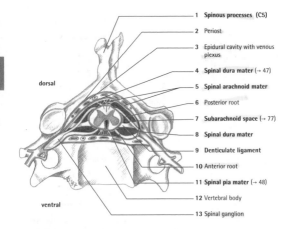

dorsal

ventral

1 **Spinous processes** (C5)

2 Periost

3 Epidural cavity with venous plexus

4 **Spinal dura mater** (→ 47)

5 **Spinal arachnoid mater**

6 Posterior root

7 **Subarachnoid space** (→ 77)

8 **Spinal dura mater**

9 Denticulate ligament

10 Anterior root

11 **Spinal pia mater** (→ 48)

12 Vertebral body

13 Spinal ganglion

At the margin of the foramen magnum the **spinal dura mater** is closely adherent to the bone. The dural sac extends to and ends at the 1st / 2nd sacral vertebrae. The spinal root nerves which emerge out of the intervertebral foramen are also covered by the dura (and also by the arachnoid membrane). Both membranes then blend with the epi- or perineurium of the nerves. Unlike the dural organization of the cranium, the spinal canal has a genuine epidural space (within the cranium the dura and periost adhere tightly to each other!).

The **epidural space** contains internal vertebral venous plexuses embedded within fibrous, loose and fatty connective tissue (protective cushion!). The internal vertebral venous plexuses communicate via the basilar plexus and the occipital sinus with the veins of the cranium (→ 65).

Keep in mind: Usually there is no **subdural space** between the dura mater and the arachnoid membrane!!

4.10.4 Arteries of Spinal Cord

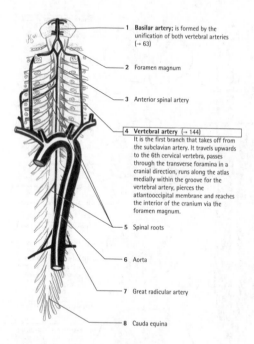

1 **Basilar artery**; is formed by the unification of both vertebral arteries (→ 63)

2 Foramen magnum

3 Anterior spinal artery

4 **Vertebral artery** (→ 144)
It is the first branch that takes off from the subclavian artery. It travels upwards to the 6th cervical vertebra, passes through the transverse foramina in a cranial direction, runs along the atlas medially within the groove for the vertebral artery, pierces the atlantooccipital membrane and reaches the interior of the cranium via the foramen magnum.

5 Spinal roots

6 Aorta

7 Great radicular artery

8 Cauda equina

4.10.5 Transverse Sections at Various Levels of Spinal Cord

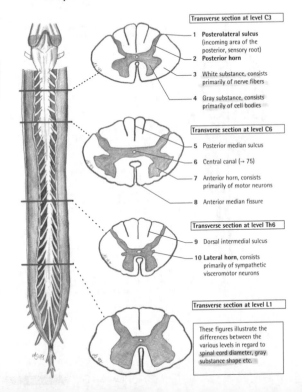

Transverse section at level C3

1 **Posterolateral sulcus**
   (incoming area of the
   posterior, sensory root)
2 **Posterior horn**

3 White substance, consists
   primarily of nerve fibers

4 Gray substance, consists
   primarily of cell bodies

Transverse section at level C6

5 Posterior median sulcus

6 Central canal (→ 75)

7 Anterior horn, consists
   primarily of motor neurons

8 Anterior median fissure

Transverse section at level Th6

9 Dorsal intermedial sulcus

10 **Lateral horn**, consists
    primarily of sympathetic
    visceromotor neurons

Transverse section at level L1

These figures illustrate the
differences between the
various levels in regard to
spinal cord diameter, gray
substance shape etc.

4.10.6 Ascending Tracts of Spinal Cord

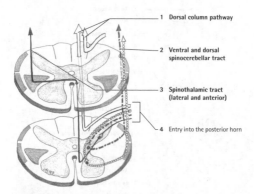

1 **Dorsal column pathway**

2 **Ventral and dorsal spinocerebellar tract**

3 **Spinothalamic tract (lateral and anterior)**

4 **Entry into the posterior horn**

Spinothalamic tract **Protopathic sensibility:** Pain, temperature	**1st neuron:** Cell body located in the spinal ganglion, the axon projects to the spinal cord – terminates (synapses) within the posterior horn on **2nd neuron**, axons from 2nd neuron cross, after traveling 1-2 segments cranially, at the gray and white commissure to the opposite side – they combine with the lateral or anterior spinothalamic tract – thalamus – postcentral gyrus.
Dorsal column pathway; **dorsal funiculus** **Epicritic sensibility** Discrimination of pressure and touch, vibration and position	**1st neuron:** Cell body located in the spinal ganglion, axons project to the spinal cord and then travel upwards towards the cranium without crossover-switchover at **2nd neuron** within the dorsal column nuclei of the medulla oblongata - axons cross at the center of the medulla oblongata to the opposite side and project as medial lemniscus to – thalamus – cerebral cortex. The **dorsal funiculus** consists of the **funiculus gracilis** (located medially) and **cuneatus** (located laterally).
Ventral and dorsal **spinocerebellar tract** **Unconscious deep tissue** **sensibility**	**1st neuron:** Cell body located in the spinal ganglion, axon projects to the spinal cord and terminates (synapses) on **2nd neuron** – axon of the 2nd neuron can take two alternative routes: a) cross to the opposite side, travel to the medulla oblongata and pons, and then enter the cerebellum via the superior cerebellar peduncles (ventral spinocerebellar tract); b) project uncrossed via the ipsilateral inferior cerebellar peduncles to the cerebellum (dorsal spinocerebellar tract).

4.11 Viscerocranium

4.11.1 Cranial Bones

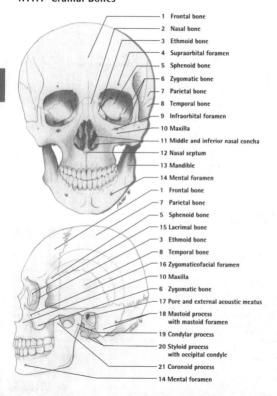

1 Frontal bone
2 Nasal bone
3 Ethmoid bone
4 Supraorbital foramen
5 Sphenoid bone
6 Zygomatic bone
7 Parietal bone
8 Temporal bone
9 Infraorbital foramen
10 Maxilla
11 Middle and inferior nasal concha
12 Nasal septum
13 Mandible
14 Mental foramen

1 Frontal bone
7 Parietal bone
5 Sphenoid bone
15 Lacrimal bone
3 Ethmoid bone
8 Temporal bone
16 Zygomaticofacial foramen
10 Maxilla
6 Zygomatic bone
17 Pore and external acoustic meatus
18 Mastoid process
   with mastoid foramen
19 Condylar process
20 Styloid process
   with occipital condyle
21 Coronoid process
14 Mental foramen

4.11.2 Paranasal Sinuses, Lacrimal Passage

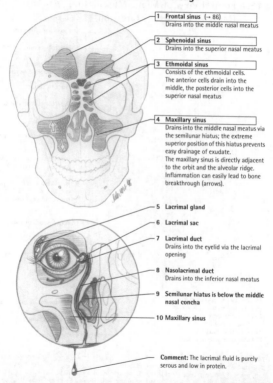

1 Frontal sinus (→ 86)
Drains into the middle nasal meatus

2 Sphenoidal sinus
Drains into the superior nasal meatus

3 Ethmoidal sinus
Consists of the ethmoidal cells.
The anterior cells drain into the middle, the posterior cells into the superior nasal meatus

4 Maxillary sinus
Drains into the middle nasal meatus via the semilunar hiatus; the extreme superior position of this hiatus prevents easy drainage of exudate.
The maxillary sinus is directly adjacent to the orbit and the alveolar ridge. Inflammation can easily lead to bone breakthrough (arrows).

5 Lacrimal gland

6 Lacrimal sac

7 Lacrimal duct
Drains into the eyelid via the lacrimal opening

8 Nasolacrimal duct
Drains into the inferior nasal meatus

9 Semilunar hiatus is below the middle nasal concha

10 Maxillary sinus

Comment: The lacrimal fluid is purely serous and low in protein.

4.12 Muscles

4.12.1 Frontal View of Mimic Musculature (Muscles of Facial Expression)

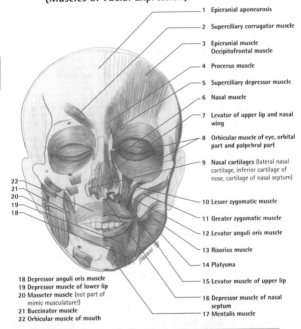

1 Epicranial aponeurosis

2 Superciliary corrugator muscle

3 Epicranial muscle
Occipitofrontal muscle

4 Procerus muscle

5 Superciliary depressor muscle

6 Nasal muscle

7 Levator of upper lip and nasal wing

8 Orbicular muscle of eye, orbital part and palpebral part

9 Nasal cartilages (lateral nasal cartilage, inferior cartilage of nose, cartilage of nasal septum)

10 Lesser zygomatic muscle

11 Greater zygomatic muscle

12 Levator anguli oris muscle

13 Risorius muscle

14 Platysma

15 Levator muscle of upper lip

16 Depressor muscle of nasal septum

17 Mentalis muscle

18 Depressor anguli oris muscle
19 Depressor muscle of lower lip
20 Masseter muscle (not part of mimic musculature!)
21 Buccinator muscle
22 Orbicular muscle of mouth

The **mimic musculature** (the masseter muscle is not a member of this group!) is exclusively innervated by the **facial nerve** (→ 103) It **does not act on joints.** The facial muscles are arranged mainly around body openings. Facial expression seems to be an inborn trait in humans. Even apes use the same nonverbal signals. Facial expression mirrors the momentary inner emotional state and is hard to suppress.

4.12.2 Lateral View of Mimic Musculature, Exiting Branches of Facial Nerve

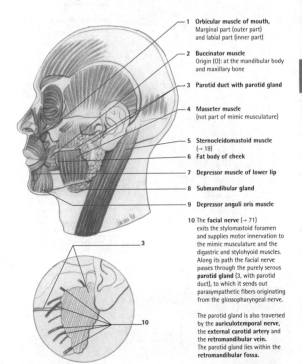

1 **Orbicular muscle of mouth,**
Marginal part (outer part)
and labial part (inner part)

2 **Buccinator muscle**
Origin (O): at the mandibular body
and maxillary bone

3 **Parotid duct with parotid gland**

4 **Masseter muscle**
(not part of mimic musculature)

5 **Sternocleidomastoid muscle**
(→ 19)

6 **Fat body of cheek**

7 **Depressor muscle of lower lip**

8 **Submandibular gland**

9 **Depressor anguli oris muscle**

10 The **facial nerve** (→ 71)
exits the stylomastoid foramen
and supplies motor innervation to
the mimic musculature and the
digastric and stylohyoid muscles.
Along its path the facial nerve
passes through the purely serous
parotid gland (3, with parotid
duct), to which it sends out
parasympathetic fibers originating
from the glossopharyngeal nerve.

The parotid gland is also traversed
by the **auriculotemporal nerve,**
the **external carotid artery** and
the **retromandibular vein.**
The parotid gland lies within the
retromandibular fossa.

4.12.3 Masticatory Musculature

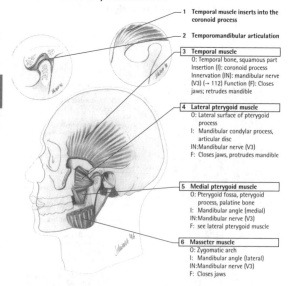

1 Temporal muscle inserts into the coronoid process

2 Temporomandibular articulation

3 Temporal muscle
O: Temporal bone, squamous part
Insertion (I): coronoid process
Innervation (IN): mandibular nerve (V3) (→ 112) Function (F): Closes jaws; retrudes mandible

4 Lateral pterygoid muscle
O: Lateral surface of pterygoid process
I: Mandibular condylar process, articular disc
IN: Mandibular nerve (V3)
F: Closes jaws, protrudes mandible

5 Medial pterygoid muscle
O: Pterygoid fossa, pterygoid process, palatine bone
I: Mandibular angle (medial)
IN: Mandibular nerve (V3)
F: see lateral pterygoid muscle

6 Masseter muscle
O: Zygomatic arch
I: Mandibular angle (lateral)
IN: Mandibular nerve (V3)
F: Closes jaws

All four masticatory muscles are innervated by branches of the mandibular nerve. Their function is to close the jaws. There are however authors, who believe pterygoid muscles also function as jaw openers.

Clinical Information

Initial signs of a progressing tetanus infection are tonic spasms of the jaw and tongue muscles. Death rigor begins within the area of the masseter muscle.

4.12.4 Floor of Mouth (Schematic)

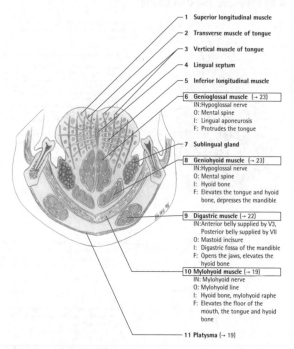

1 **Superior longitudinal muscle**

2 **Transverse muscle of tongue**

3 **Vertical muscle of tongue**

4 **Lingual septum**

5 **Inferior longitudinal muscle**

6 **Genioglossal muscle** (→ 23)
IN: Hypoglossal nerve
O: Mental spine
I: Lingual aponeurosis
F: Protrudes the tongue

7 **Sublingual gland**

8 **Geniohyoid muscle** (→ 23)
IN: Hypoglossal nerve
O: Mental spine
I: Hyoid bone
F: Elevates the tongue and hyoid bone, depresses the mandible

9 **Digastric muscle** (→ 22)
IN: Anterior belly supplied by V3, Posterior belly supplied by VII
O: Mastoid incisure
I: Digastric fossa of the mandible
F: Opens the jaws, elevates the hyoid bone

10 **Mylohyoid muscle** (→ 19)
IN: Mylohyoid nerve
O: Mylohyoid line
I: Hyoid bone, mylohyoid raphe
F: Elevates the floor of the mouth, the tongue and hyoid bone

11 **Platysma** (→ 19)

4.13 Tongue, Oral Cavity

4.13.1 Tongue and Oral Cavity

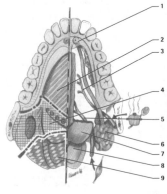

1 **Sublingual caruncle**, receives drainage from the submandibular and major sublingual duct.

2 **Area innervated by the lingual nerve** (→ 71)

3 **Lingual nerve (V3);** directs gustatory fibers originating from the anterior two-third of the tongue, via the cord of tympanum to the facial nerve. This area receives sensory innervation (including teeth and oral vestibule) from the trigeminal nerve (V).

4 Area receiving sensory innervation from the **glossopharyngeal nerve (IX);** also shown are the vallate papillae, the terminal sulcus and the palatine tonsil.

5 **Hypoglossal nerve (XII);** provides motor innervation of the tongue and infrahyoid musculature.

6 **Submandibular gland**

7 **Lingual tonsil**

8 **Epiglottis** (→ 27)

9 Receptive area receiving sensory innervation from the **vagus nerve (X)**

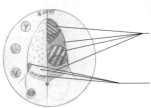

10 The four basic **taste sensations** of the tongue from the anterior tip to the posterior base are: sweet, salty, sour, bitter. All other taste sensations and flavors are perceived through the olfactory organ.

The papillae of the tongue are located in the following areas:
- the **filiform papillae** (touch sensation) and the **fungiform papillae** (taste sensation) at the dorsum of the tongue
- the **foliate papillae** (taste sensation) at the back margin of the tongue
- the **vallate papillae** (taste sensation) anterior to the terminal sulcus

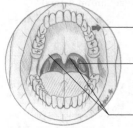

The parotid duct drains into the oral vestibule, across from the 2nd molar at the parotid papillae. The parotid duct travels below the zygomatic arch, over the masseter muscle and pierces the buccinator muscle.

Palatine tonsil
The tonsils encircle the pharyngeal entrance (Waldeyer's tonsillar ring). They consist of the four large tonsils (the palatine, lingual and pharyngeal tonsils), the tubal tonsil and the so-called lateral pharyngeal bands at the lateral pharyngeal wall.

Palatoglossal (front) and
Palatopharyngeal arch (back)

4.14 Extracranial Vasculature, Nerves, Lymphatic Vessels

4.14.1 Arteries of Head (External Carotid Artery)

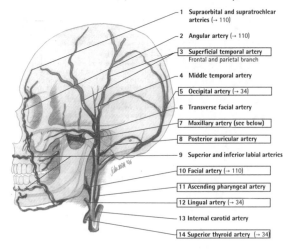

1 Supraorbital and supratrochlear arteries (→ 110)

2 Angular artery (→ 110)

3 Superficial temporal artery
Frontal and parietal branch

4 Middle temporal artery

5 Occipital artery (→ 34)

6 Transverse facial artery

7 Maxillary artery (see below)

8 Posterior auricular artery

9 Superior and inferior labial arteries

10 Facial artery (→ 110)

11 Ascending pharyngeal artery

12 Lingual artery (→ 34)

13 Internal carotid artery

14 Superior thyroid artery (→ 34)

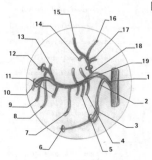

1 **Maxillary artery**
(projects interiorly between the temporal and lateral pterygoid muscle)
2. Inferior alveolar artery
3. Mylohyoid branch
4. Masseteric artery
5. Pterygoid branches
6. Mental artery
7. Buccal artery
8. Descending palatine artery
9. Superior anterior alveolar artery
10. Infraorbital artery
11. Superior posterior alveolar artery
12. Sphenopalatine artery
13. Artery of pterygoid canal
14. Orbital branch
15. Frontal branch
16. Parietal branch
17. Superior tympanic artery, petrosal branch
18. Middle meningeal artery
(surrounded by the root branches of the auriculotemporal nerve)
19. Deep temporal arteries

4.14.2 Vessels and Nerves of Head

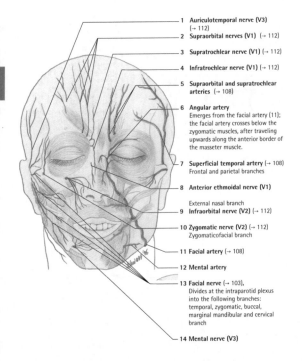

1 **Auriculotemporal nerve (V3)**
(→ 112)

2 **Supraorbital nerves (V1)** (→ 112)

3 **Supratrochlear nerve (V1)** (→ 112)

4 **Infratrochlear nerve (V1)** (→ 112)

5 **Supraorbital and supratrochlear arteries** (→ 108)

6 **Angular artery**
Emerges from the facial artery (11); the facial artery crosses below the zygomatic muscles, after traveling upwards along the anterior border of the masseter muscle.

7 **Superficial temporal artery** (→ 108)
Frontal and parietal branches

8 **Anterior ethmoidal nerve (V1)**

External nasal branch

9 **Infraorbital nerve (V2)** (→ 112)

10 **Zygomatic nerve (V2)** (→ 112)
Zygomaticofacial branch

11 **Facial artery** (→ 108)

12 **Mental artery**

13 **Facial nerve** (→ 103),
Divides at the intraparotid plexus into the following branches: temporal, zygomatic, buccal, marginal mandibular and cervical branch

14 **Mental nerve (V3)**

4.14.3 Trigeminal Nerve In Situ

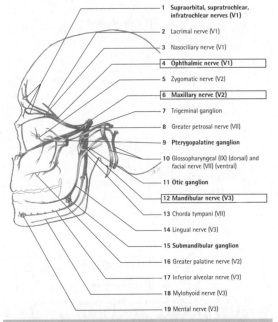

1 Supraorbital, supratrochlear, infratrochlear nerves (V1)

2 Lacrimal nerve (V1)

3 Nasociliary nerve (V1)

4 Ophthalmic nerve (V1)

5 Zygomatic nerve (V2)

6 Maxillary nerve (V2)

7 Trigeminal ganglion

8 Greater petrosal nerve (VII)

9 Pterygopalatine ganglion

10 Glossopharyngeal (IX) (dorsal) and facial nerve (VII) (ventral)

11 Otic ganglion

12 Mandibular nerve (V3)

13 Chorda tympani (VII)

14 Lingual nerve (V3)

15 Submandibular ganglion

16 Greater palatine nerve (V2)

17 Inferior alveolar nerve (V3)

18 Mylohyoid nerve (V3)

19 Mental nerve (V3)

The trigeminal nerve emerges from the brain at the lateral area of the pons (→ 70)

It is primarily a **sensory nerve**, that carries fibers from the face, nasal cavity, paranasal sinuses, oral cavity and teeth.

Fibers from its somewhat smaller **motor root**, project jointly with the mandibular nerve to the muscles of mastication and to the floor of the mouth.

4.14.4 Trigeminal Nerve, Branches (Schematic)

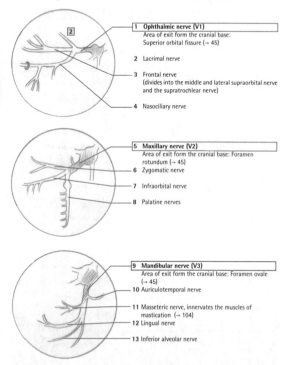

1 Ophthalmic nerve (V1)
Area of exit form the cranial base:
Superior orbital fissure (→ 45)

2 Lacrimal nerve

3 Frontal nerve
(divides into the middle and lateral supraorbital nerve
and the supratrochlear nerve)

4 Nasociliary nerve

5 Maxillary nerve (V2)
Area of exit form the cranial base: Foramen
rotundum (→ 45)

6 Zygomatic nerve

7 Infraorbital nerve

8 Palatine nerves

9 Mandibular nerve (V3)
Area of exit form the cranial base: Foramen ovale
(→ 45)

10 Auriculotemporal nerve

11 Masseteric nerve, innervates the muscles of
mastication (→ 104)

12 Lingual nerve

13 Inferior alveolar nerve

4.14.5 Sensory Innervation of Face

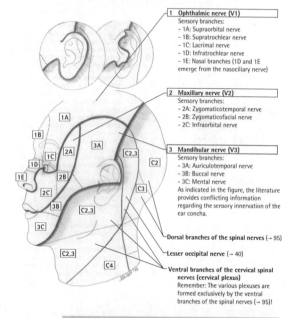

1 Ophthalmic nerve (V1)
Sensory branches:
- 1A: Supraorbital nerve
- 1B: Supratrochlear nerve
- 1C: Lacrimal nerve
- 1D: Infratrochlear nerve
- 1E: Nasal branches (1D and 1E emerge from the nasociliary nerve)

2 Maxillary nerve (V2)
Sensory branches:
- 2A: Zygomaticotemporal nerve
- 2B: Zygomaticofacial nerve
- 2C: Infraorbital nerve

3 Mandibular nerve (V3)
Sensory branches:
- 3A: Auriculotemporal nerve
- 3B: Buccal nerve
- 3C: Mental nerve
As indicated in the figure, the literature provides conflicting information regarding the sensory innervation of the ear concha.

Dorsal branches of the spinal nerves (→ 95)

Lesser occipital nerve (→ 40)

Ventral branches of the cervical spinal nerves (cervical plexus)
Remember: The various plexuses are formed exclusively by the ventral branches of the spinal nerves (→ 95)!

Clinical Information

Shown in the figure are the pressure points of the trigeminal nerve, located above the supraorbital, infraorbital and mental foramina. Essential trigeminal neuralgia is characterized by hyperpathia and hypersensibility.

Painful sensations for example occur when applying pressure on the exiting areas of the nerves.

4.14.6 Parasympathetic Ganglia of Head

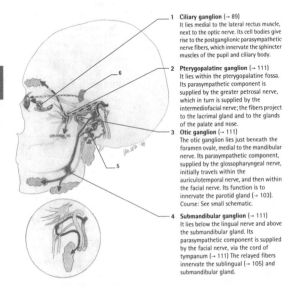

1   **Ciliary ganglion** (→ 89)
    It lies medial to the lateral rectus muscle, next to the optic nerve. Its cell bodies give rise to the postganglionic parasympathetic nerve fibers, which innervate the sphincter muscles of the pupil and ciliary body.

2   **Pterygopalatine ganglion** (→ 111)
    It lies within the pterygopalatine fossa. Its parasympathetic component is supplied by the greater petrosal nerve, which in turn is supplied by the intermediofacial nerve; the fibers project to the lacrimal gland and to the glands of the palate and nose.

3   **Otic ganglion** (→ 111)
    The otic ganglion lies just beneath the foramen ovale, medial to the mandibular nerve. Its parasympathetic component, supplied by the glossopharyngeal nerve, initially travels within the auriculotemporal nerve, and then within the facial nerve. Its function is to innervate the parotid gland (→ 103).
    Course: See small schematic.

4   **Submandibular ganglion** (→ 111)
    It lies below the lingual nerve and above the submandibular gland. Its parasympathetic component is supplied by the facial nerve, via the cord of tympanum (→ 111) The relayed fibers innervate the sublingual (→ 105) and submandibular gland.

Only the parasympathetic, preganglionic fibers are switched over in the parasympathetic ganglia of the head and then leave the ganglia as postganglionic fibers. The sympathetic and sensory fiber tracts pass without switchover through these ganglia, to innervate their target organs.
The structure labeled as "5" in the figure represents the superior cervical sympathetic ganglion of the sympathetic trunk. The number "6" depicts the sympathetic internal carotid plexus.

4.14.7 Veins of Head

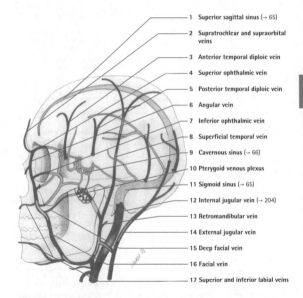

1 Superior sagittal sinus (→ 65)
2 Supratrochlear and supraorbital veins
3 Anterior temporal diploic vein
4 Superior ophthalmic vein
5 Posterior temporal diploic vein
6 Angular vein
7 Inferior ophthalmic vein
8 Superficial temporal vein
9 Cavernous sinus (→ 66)
10 Pterygoid venous plexus
11 Sigmoid sinus (→ 65)
12 Internal jugular vein (→ 204)
13 Retromandibular vein
14 External jugular vein
15 Deep facial vein
16 Facial vein
17 Superior and inferior labial veins

Clinical Information

Under certain circumstances the facial veins can transmit facial infections into the interior of the cranium (e.g. as may happen in the case of a furuncle). This is due to the fact that the blood in areas above the corners of the mouth is routed directly via the inferior and superior ophthalmic veins into the brain sinus (→ 65).

4.14.8 Lymphatic Vessels of Head and Neck

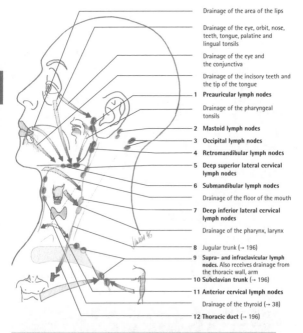

Drainage of the area of the lips

Drainage of the eye, orbit, nose, teeth, tongue, palatine and lingual tonsils

Drainage of the eye and the conjunctiva

Drainage of the incisory teeth and the tip of the tongue

1 **Preauricular lymph nodes**

Drainage of the pharyngeal tonsils

2 **Mastoid lymph nodes**

3 **Occipital lymph nodes**

4 **Retromandibular lymph nodes**

5 **Deep superior lateral cervical lymph nodes**

6 **Submandibular lymph nodes**

Drainage of the floor of the mouth

7 **Deep inferior lateral cervical lymph nodes**

Drainage of the pharynx, larynx

8 **Jugular trunk** (→ 196)

9 **Supra- and infraclavicular lymph nodes.** Also receives drainage from the thoracic wall, arm

10 **Subclavian trunk** (→ 196)

11 **Anterior cervical lymph nodes**

Drainage of the thyroid (→ 38)

12 **Thoracic duct** (→ 196)

Arrows in the figure with darkened heads indicate deep-lying drainage areas. Drainage of the **lymph of the head** into the venous angles (between the jugular and subclavian veins) occurs on the left side via the thoracic duct and on the right side via the right lymphatic duct.

Clinic: Palpable **enlargement of the lymph node** often indicates the prodromal stage of a disease.

5.   Ventral Wall of Trunk

5.1   General Facts

5.1.1   Regions of the Ventral Wall of Trunk

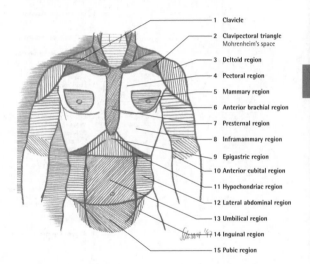

1 Clavicle

2 Clavipectoral triangle
   Mohrenheim's space

3 Deltoid region

4 Pectoral region

5 Mammary region

6 Anterior brachial region

7 Presternal region

8 Inframammary region

9 Epigastric region

10 Anterior cubital region

11 Hypochondriac region

12 Lateral abdominal region

13 Umbilical region

14 Inguinal region

15 Pubic region

5.2 Muscles, Vessels

5.2.1 Muscles of Ventral Wall of Trunk I

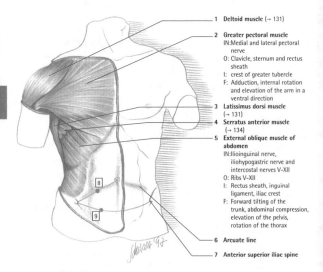

1 **Deltoid muscle** (→ 131)

2 **Greater pectoral muscle**
IN: Medial and lateral pectoral nerve
O: Clavicle, sternum and rectus sheath
I: crest of greater tubercle
F: Adduction, internal rotation and elevation of the arm in a ventral direction

3 **Latissimus dorsi muscle** (→ 131)

4 **Serratus anterior muscle** (→ 134)

5 **External oblique muscle of abdomen**
IN: Ilioinguinal nerve, iliohypogastric nerve and intercostal nerves V–XII
O: Ribs V–XII
I: Rectus sheath, inguinal ligament, iliac crest
F: Forward tilting of the trunk, abdominal compression, elevation of the pelvis, rotation of the thorax

6 **Arcuate line**

7 **Anterior superior iliac spine**

Clinical Information

Even today, acute appendicitis is a disease hard to diagnose. Two established pain pressure points are described here: No. 8 is known as **McBurney's point**. It lies in the middle of an imaginary line between the anterior superior iliac spine and the navel. No. 9 is known as **Lanz's point** and is located at the first third of a line connecting the two spines.

Additional symptoms of appendicitis are: Nausea, vomiting, abdominal pain and local tension of the abdominal muscles.

5.2.2 Muscles of Ventral Wall of Trunk II

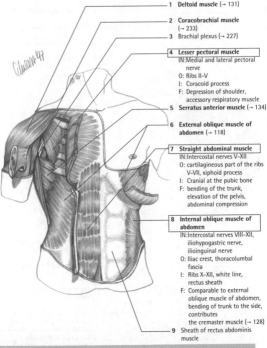

1 Deltoid muscle (→ 131)

2 Coracobrachial muscle
(→ 233)

3 Brachial plexus (→ 227)

4 Lesser pectoral muscle
IN: Medial and lateral pectoral
nerve
O: Ribs II-V
I: Coracoid process
F: Depression of shoulder,
accessory respiratory muscle

5 Serratus anterior muscle (→ 134)

**6 External oblique muscle of
abdomen** (→ 118)

7 Straight abdominal muscle
IN: Intercostal nerves V-XII
O: cartilagineous part of the ribs
V-VII, xiphoid process
I: Cranial at the pubic bone
F: bending of the trunk,
elevation of the pelvis,
abdominal compression

**8 Internal oblique muscle of
abdomen**
IN: Intercostal nerves VIII-XII,
iliohypogastric nerve,
ilioinguinal nerve
O: Iliac crest, thoracolumbar
fascia
I: Ribs X-XII, white line,
rectus sheath
F: Comparable to external
oblique muscle of abdomen,
bending of trunk to the side,
contributes
the cremaster muscle (→ 128)

**9 Sheath of rectus abdominis
muscle**

Clearly visible are the 3 – 4 **tendinous intersections**, which together with the medial border of the
muscle, adhere to the anterior lamina of the rectus sheath. Unlike the depiction in this figure, they
are usually only located above the arcuate line.

5.2.3   Rectus Sheath

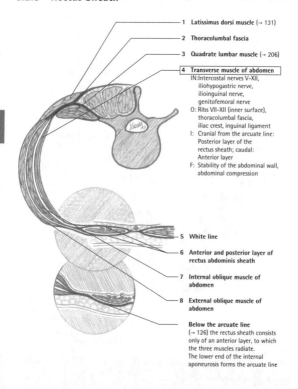

1 Latissimus dorsi muscle (→ 131)

2 Thoracolumbal fascia

3 Quadrate lumbar muscle (→ 206)

4 Transverse muscle of abdomen
   IN: Intercostal nerves V-XII,
       iliohypogastric nerve,
       ilioinguinal nerve,
       genitofemoral nerve
   O: Ribs VII-XII (inner surface),
      thoracolumbal fascia,
      iliac crest, inguinal ligament
   I: Cranial from the arcuate line:
      Posterior layer of the
      rectus sheath; caudal:
      Anterior layer
   F: Stability of the abdominal wall,
      abdominal compression

5 White line

6 Anterior and posterior layer of
   rectus abdominis sheath

7 Internal oblique muscle of
   abdomen

8 External oblique muscle of
   abdomen

Below the arcuate line
(→ 126) the rectus sheath consists
only of an anterior layer, to which
the three muscles radiate.
The lower end of the internal
aponeurosis forms the arcuate line

5.2.4 Muscles of Abdomen (Schematic)

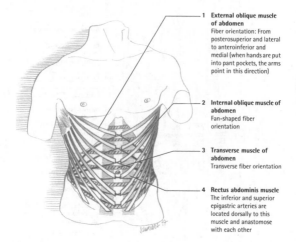

1 **External oblique muscle of abdomen**
Fiber orientation: From posterosuperior and lateral to anteroinferior and medial (when hands are put into pant pockets, the arms point in this direction)

2 **Internal oblique muscle of abdomen**
Fan-shaped fiber orientation

3 **Transverse muscle of abdomen**
Transverse fiber orientation

4 **Rectus abdominis muscle**
The inferior and superior epigastric arteries are located dorsally to this muscle and anastomose with each other

5.2.5 Sagittal Section of Thoracic Wall

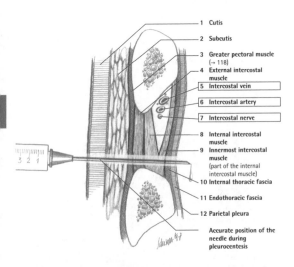

1 Cutis
2 Subcutis
3 Greater pectoral muscle (→ 118)
4 External intercostal muscle
5 Intercostal vein
6 Intercostal artery
7 Intercostal nerve
8 Internal intercostal muscle
9 Innermost intercostal muscle (part of the internal intercostal muscle)
10 Internal thoracic fascia
11 Endothoracic fascia
12 Parietal pleura
Accurate position of the needle during pleurocentesis

Clinical Information

Die intercostal vessels extend within the costal sulcus to the midaxillary line. Pleurocentesis is always performed at the **superior border** of a rib. Several authors recommended the following insertion sites:

- 5th to 7th intercostal space: Puncture dorsal to the midaxillary line
- 4th to 6th intercostal space: Puncture at the midaxillary line
- 2nd to 3rd intercostal space: Puncture at the midclavicular line

5.2.6 Respiratory Mechanics

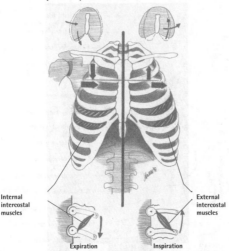

Internal intercostal muscles

External intercostal muscles

Expiration   **Inspiration**

Shown in the figure above are the events, which take place during expiration (left) and inspiration (right). Always keep in mind however, that the primary muscle of respiration is the **diaphragm** (→ 155) but not the intercostal muscle system. During expiration the ribs move downward and rotate slightly, due to the contraction of the **internal intercostal muscles**.

The internal intercostal muscles run from the superior border of a rib to the inferior border of the next higher rib. The fiber orientation of these muscles is shown in the figure. This mode of attachment results in a **downward pull**, since the lever arm of the upper rib is significantly longer than that of the lower rib.

During **inspiration** the **external intercostal muscles** elevate and rotate the ribs and thereby expand the thoracic cavity. Due to the fiber orientation of the external intercostal muscles, the lever arm configuration is exactly opposite to expiration and this in turn leads to an **upward pull**. The external intercostal muscles run from the superior border of a rib to the superior border of the next rib below. The fiber orientation of these muscles is shown in the figure.

5.2.7 Veins of Ventral Trunk Wall

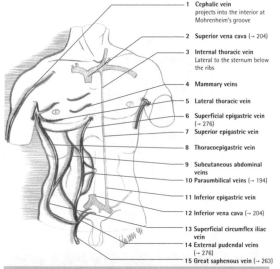

1 Cephalic vein
   projects into the interior at
   Mohrenheim's groove

2 Superior vena cava (→ 204)

3 Internal thoracic vein
   Lateral to the sternum below
   the ribs

4 Mammary veins

5 Lateral thoracic vein

6 Superficial epigastric vein
   (→ 276)

7 Superior epigastric vein

8 Thoracoepigastric vein

9 Subcutaneous abdominal
   veins

10 Paraumbilical veins (→ 194)

11 Inferior epigastric vein

12 Inferior vena cava (→ 204)

13 Superficial circumflex iliac
   vein

14 External pudendal veins
   (→ 276)

15 Great saphenous vein (→ 263)

Clinical Information

The veins of the ventral trunk wall are clinically important since they are involved in the
development of collateral circulation (**portocaval anastomosis**). These are alternative routes
of blood supply that originate from the portal vein of the liver and end in the superior and
inferior vena cava. They bypass the liver during impairment of portal flow. The following routes
are available:

- via the veins of the lesser curvature of the stomach and the esophagus to the azygous vein and
  then to the superior vena cava.
- via the rectal veins and the rectal venous plexus to the internal iliac vein and then to the
  inferior vena cava.
- via the paraumbilical veins to veins of the skin, for example to the superficial epigastric vein
  (looks very impressive and is also known as "Medusa's head") and then via the femoral vein to
  the inferior vena cava.

5.2.8 Dorsal View of Anterior Abdominal Wall

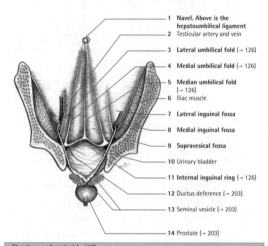

1 Navel. Above is the hepatoumbilical ligament
2 Testicular artery and vein
3 Lateral umbilical fold (→ 126)
4 Medial umbilical fold (→ 126)
5 Median umbilical fold (→ 126)
6 Iliac muscle
7 Lateral inguinal fossa
8 Medial inguinal fossa
9 Supravesical fossa
10 Urinary bladder
11 Internal inguinal ring (→ 126)
12 Ductus deference (→ 203)
13 Seminal vesicle (→ 203)
14 Prostate (→ 203)

Structures and content (→ 127)	
Round ligament of liver (→ 180)	Umbilical vein
Lateral umbilical fold (→ 126)	Inferior epigastric blood vessels
Medial umbilical fold (→ 126)	Umbilical artery (→ 157)
Median umbilical fold (→ 126)	Median umbilical ligament
	(obliterated urachal canal)

Clinical Information

The internal inguinal ring is located within the lateral inguinal fossa. **Indirect inguinal hernia** occurs when the hernial content (e.g. section of the intestine) enters through the internal inguinal ring. **Direct inguinal hernia** occurs, when the hernial content enters the inguinal canal directly through the musculature at the medial inguinal fossa (→ 129).

5.2.9 Inguinal Canal

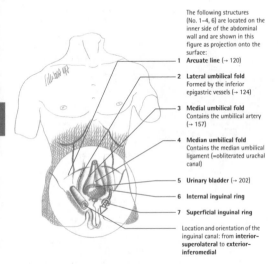

The following structures (No. 1–4, 6) are located on the inner side of the abdominal wall and are shown in this figure as projection onto the surface:

1 **Arcuate line** (→ 120)

2 **Lateral umbilical fold**
Formed by the inferior epigastric vessels (→ 124)

3 **Medial umbilical fold**
Contains the umbilical artery (→ 157)

4 **Median umbilical fold**
Contains the median umbilical ligament (=obliterated urachal canal)

5 **Urinary bladder** (→ 202)

6 **Internal inguinal ring**

7 **Superficial inguinal ring**

Location and orientation of the inguinal canal: from **interior-superolateral** to **exterior-inferomedial**

Boundaries of the inguinal canal	
Ventral	Superficial abdominal fascia, aponeurosis of the external oblique abdominal muscle
Dorsal	Peritoneum, transversal fascia, inguinal falx, interfoveolar ligament
Cranial	Internal oblique muscle of abdomen, transverse muscle of abdomen
Caudal	Inguinal ligament, reflex ligament of Gimbernat

5.2.10 Content of Inguinal Canal

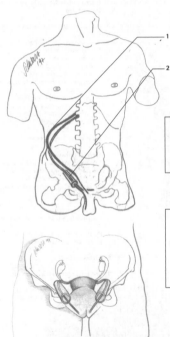

1 **Iliohypogastric nerve** (→ 207)
Does not traverse
the inguinal canal

2 **Ilioinguinal nerve** (→ 207)
Traverses the inguinal canal.

In males the following structures
pass through the inguinal canal:
- **Spermatic cord** (→ 128)
- **Genital branch of
  genitofemoral nerve**
- **Ilioinguinal nerve**

In females the following structures
pass through the inguinal canal:
- **Artery of round ligament of
  uterus**
- **Round ligament of uterus**
  (→ 212) **(see figure)**
- **Genital branch of
  genitofemoral nerve**
- **Ilioinguinal nerve**

This figure shows clearly, that the
inguinal canal runs **above** the
inguinal ligament.

5.2.11 Spermatic Cord, Coverings of Testis

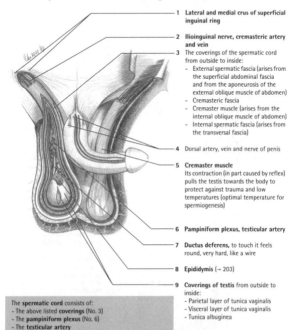

1 **Lateral and medial crus of superficial inguinal ring**

2 **Ilioinguinal nerve, cremasteric artery and vein**

3 The coverings of the spermatic cord from outside to inside:
  - External spermatic fascia (arises from the superficial abdominal fascia and from the aponeurosis of the external oblique muscle of abdomen)
  - Cremasteric fascia
  - Cremaster muscle (arises from the internal oblique muscle of abdomen)
  - Internal spermatic fascia (arises from the transversal fascia)

4 **Dorsal artery, vein and nerve of penis**

5 **Cremaster muscle**
  Its contraction (in part caused by reflex) pulls the testis towards the body to protect against trauma and low temperatures (optimal temperature for spermiogenesis)

6 **Pampiniform plexus, testicular artery**

7 **Ductus deferens,** to touch it feels round, very hard, like a wire

8 **Epididymis** (→ 203)

9 **Coverings of testis** from outside to inside:
  - Parietal layer of tunica vaginalis
  - Visceral layer of tunica vaginalis
  - Tunica albuginea

The **spermatic cord** consists of:
- The above listed **coverings** (No. 3)
- The **pampiniform plexus** (No. 6)
- The **testicular artery**
- The **ductus deferens** (No. 7)

The layers **that surround the testis** develop within the embryonal descensus testis from fascial outpouchings of the three ventral trunk wall muscles (external oblique, internal oblique and transverse muscle of abdomen) and from outpouchings of the parietal and visceral peritoneum!

5.2.12 Hernias

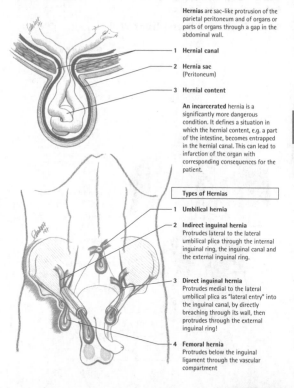

Hernias are sac-like protrusion of the parietal peritoneum and of organs or parts of organs through a gap in the abdominal wall.

1 **Hernial canal**

2 **Hernia sac**
(Peritoneum)

3 **Hernial content**

An incarcerated hernia is a significantly more dangerous condition. It defines a situation in which the hernial content, e.g. a part of the intestine, becomes entrapped in the hernial canal. This can lead to infarction of the organ with corresponding consequences for the patient.

Types of Hernias

1 **Umbilical hernia**

2 **Indirect inguinal hernia**
Protrudes lateral to the lateral umbilical plica through the internal inguinal ring, the inguinal canal and the external inguinal ring.

3 **Direct inguinal hernia**
Protrudes medial to the lateral umbilical plica as "lateral entry" into the inguinal canal, by directly breaching through its wall, then protrudes through the external inguinal ring!

4 **Femoral hernia**
Protrudes below the inguinal ligament through the vascular compartment

6. Dorsal Trunk Wall

6.1 General Facts

6.1.1 Regions of Dorsal Trunk Wall

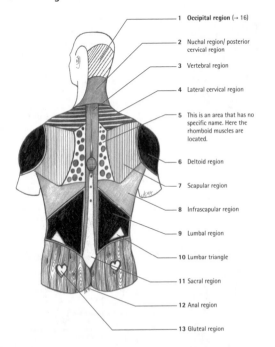

1 **Occipital region** (→ 16)

2 Nuchal region/ posterior cervical region

3 Vertebral region

4 Lateral cervical region

5 This is an area that has no specific name. Here the rhomboid muscles are located.

6 Deltoid region

7 Scapular region

8 Infrascapular region

9 Lumbal region

10 Lumbar triangle

11 Sacral region

12 Anal region

13 Gluteal region

6.2 Muscles, Vessels, Nerves

6.2.1 Superficial Musculature of Back

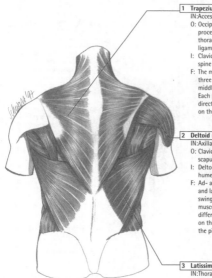

1 Trapezius muscle
IN: Accessory nerve (XI)
O: Occipital bone, spinous processes of the cervical and thoracic vertebrae, nuchal ligament
I: Clavicle, acromion, scapular spine
F: The muscle is divided into three parts: The superior, middle and inferior part. Each part has a different direction of pull, depending on the lever it acts upon

2 Deltoid muscle
IN: Axillary nerve
O: Clavicle, acromion and scapular spine
I: Deltoid tuberosity of humerus
F: Ad- and Abduction, medial and lateral rotation, swinging of the arm; the muscular subdivisions exert different actions, depending on their location relative to the pivoting point

3 Latissimus dorsi muscle
IN: Thoracodorsal nerve
O: Spinous processes of thoracic and lumbar vertebrae, sacral bone, iliac crest, thoracolumbar fascia
I: Minor tubercular crest
F: Adduction, medial rotation of humerus

6.2.2 Deep Musculature of Back I

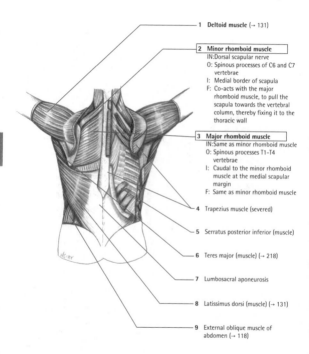

1 **Deltoid muscle** (→ 131)

2 **Minor rhomboid muscle**
   IN: Dorsal scapular nerve
   O: Spinous processes of C6 and C7 vertebrae
   I: Medial border of scapula
   F: Co-acts with the major rhomboid muscle, to pull the scapula towards the vertebral column, thereby fixing it to the thoracic wall

3 **Major rhomboid muscle**
   IN: Same as minor rhomboid muscle
   O: Spinous processes T1-T4 vertebrae
   I: Caudal to the minor rhomboid muscle at the medial scapular margin
   F: Same as minor rhomboid muscle

4 **Trapezius muscle** (severed)

5 **Serratus posterior inferior** (muscle)

6 **Teres major** (muscle) (→ 218)

7 **Lumbosacral aponeurosis**

8 **Latissimus dorsi** (muscle) (→ 131)

9 **External oblique muscle of abdomen** (→ 118)

6.2.3 Deep Musculature of Back II

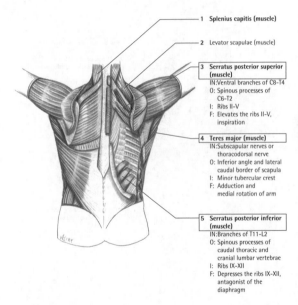

1 **Splenius capitis (muscle)**

2 Levator scapulae (muscle)

3 **Serratus posterior superior (muscle)**
IN:Ventral branches of C8-T4
O: Spinous processes of C6-T2
I: Ribs II-V
F: Elevates the ribs II-V, inspiration

4 **Teres major (muscle)**
IN:Subscapular nerves or thoracodorsal nerve
O: Inferior angle and lateral caudal border of scapula
I: Minor tubercular crest
F: Adduction and medial rotation of arm

5 **Serratus posterior inferior (muscle)**
IN:Branches of T11-L2
O: Spinous processes of caudal thoracic and cranial lumbar vertebrae
I: Ribs IX-XII
F: Depresses the ribs IX-XII, antagonist of the diaphragm

6.2.4 Serratus Anterior (Muscle)

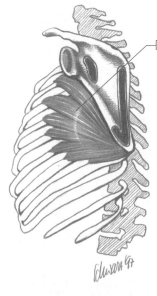

1 **Serratus anterior (muscle)**
   IN: Long thoracic nerve
   O: Ribs I–IX, divided into superior, middle and inferior part
   I: Superior angle, medial border and inferior angle of scapula; the muscle runs ventral to the subscapular muscle below the shoulder blade; seen from the back the sequence therefore would be: Scapula – subscapular muscle – serratus anterior (muscle)
   F: Rotation of the scapula (important for arm movement!), fixation of the scapula, may also act as accessory respiratory muscle, together with the rhomboid muscle it forms a muscular sling (stability of the trunk).

For better visualization, the figure shows the muscle insertions located on the ventral side of the scapula, projected onto the dorsal side of the shoulder blade. The scapula is quasi "transparent".

6.2.5 Autochthonous Musculature of Back

Those muscles of the back that are not shifted dorsally from other locations during embryonic development, but instead remain at the site of origin, are called **autochthonous back muscles.**

The muscle tracts that jointly form the **erector muscle of spine,** are innervated by the **dorsal branches of the spinal nerves** (→ 95).

The remaining back muscles are innervated by the ventral branches. The **autochthonous back muscles** support the vertebral column through contraction, which builds up pressure within the osseofibrous tunnel resulting in the stiffening of the skeletal axis. In addition it participates in all vertebral column movements.

A classification of the autochthonous back muscles follows below:

	Straight system	Angular system
Medial column	**Interspinal straight system** (→ 137) Interspinal muscles Spinal muscle Intertransverse muscle	**Transversospinal angular system** (→ 138) Rotatores muscles Multifidus muscle Semispinal muscles
Lateral column	**Intertransversal straight system** (→ 139) Longissimus muscle Iliocostalis muscle	**Spinotransversal angular system** (→ 140) Splenius muscle

6.2.6 Osseofibrous Tunnel

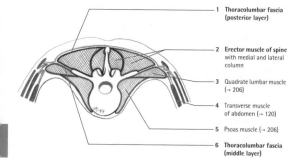

1 Thoracolumbar fascia (posterior layer)

2 Erector muscle of spine with medial and lateral column

3 Quadrate lumbar muscle (→ 206)

4 Transverse muscle of abdomen (→ 120)

5 Psoas muscle (→ 206)

6 Thoracolumbar fascia (middle layer)

Shown above is a transverse section depicting the **autochthonous back musculature** within the **osseofibrous tunnel**. This tunnel is formed by the **thoracolumbar fascia**, in conjunction with the spinous processes and ribs of the vertebrae. However, within the thoracic area of the spine, the autochthonous muscle fiber tracts are also in contact with the ribs.

The **middle layer** of the thoracolumbar fascia arises from the costal processes of the lumbar vertebrae. It extends between the ventrolateral and autochthonous musculature. From this layer the internal oblique and transverse muscle of abdomen arise.

The **posterior layer** of the thoracolumbal fascia arises from the spinous processes, the iliac crest and from the costal angles. It projects cranially and continues into the nuchal fascia. The posterior and inferior serratus muscle and latissimus dorsi muscle project from the posterior layer to reach their particular insertion points.

6.2.7 Medial Column – Straight System

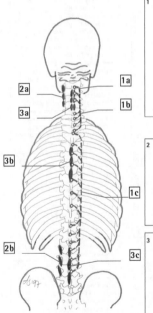

1 Spinalis (muscle) (skips at least 1 vertebra)
- Spinalis capitis (muscle) (1a): Runs from the spinous processes of the caudal cervical vertebrae to the superior nuchal line
- Spinalis cervicis (muscle) (1b): Runs from the spinous processes of the cranial thoracic vertebrae to the spinous processes of the medial cervical vertebrae
- Spinalis thoracis (muscle) (1c): Runs from the spinous processes of the cranial lumbar vertebrae to the spinous processes of the medial thoracic vertebrae

2 Intertransverse muscle
   Posterior cervical intertransverse muscles (2a): Run between the transverse processes of adjacent cervical vertebrae
   Medial lumbar intertransverse muscles (2b): Run from the accessory processes of a lumbar vertebra to the mamillary processes of the next lumbar vertebra below

3 Interspinal muscle
   A distinction is drawn between: **Cervical (3a), thoracic (3b)** and **lumbar (3c) interspinal muscles.**
   They run in pairs from the spinous processes of a vertebra to the spinous processes of the next adjacent vertebra without skipping a segment. These muscles are not present in the medial thoracic area of the spine

6.2.8   Medial Column – Angular System

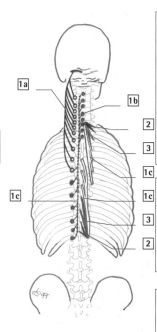

1   **Semispinalis muscle**
    (skips approximately 5 segments)
    - **Semispinalis capitis (muscle) (1a):**
      Runs from the transverse processes of
      the caudal cervical and cranial
      thoracic vertebrae to the superior and
      inferior nuchal line
    - **Semispinalis cervicis (muscle) (1b):**
      Runs from the transverse processes of
      the thoracic vertebrae to the spinous
      processes of higher lying thoracic and
      cervical vertebrae
    - **Semispinalis thoracis (muscle) (1c):**
      Runs from the transverse processes of
      the lower thoracic vertebrae to the
      spinous processes of the upper
      thoracic vertebrae; 1b and 1c cannot
      be distinguished from each other

2   **Rotatores (muscles)**
    The rotatores muscles are located in
    the cervical, thoracic and lumbar
    areas. Distinguished are:
    Rotatores brevis muscles, which run
    between adjacent vertebrae, and
    rotatores longus muscles, which skip
    1 segment each time. Originating
    from transverse processes, the
    rotatores muscles project to higher
    spinous processes

3   **Multifidi muscles**
    (skip up to 3 segments)
    The multifidi muscles run just as the
    rotatores muscles, but skip a larger
    number of vertebrae.

6.2.9 Lateral Column – Straight System

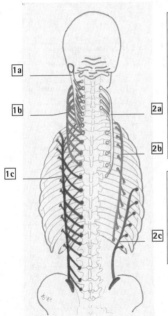

1 **Longissimus muscle (medial from 2)**
- **Longissimus capitis (muscle)** (1a):
  Projects from the transverse processes of the cervical and thoracic vertebrae to the mastoid process
- **Longissimus cervicis (muscle)** (1b): Projects from the transverse processes of the thoracic vertebrae to the transverse processes of the cervical vertebrae
- **Longissimus thoracis (muscle)** (1c): Projects from the sacral bone and transverse processes of the lumbar vertebrae to all ribs and to the transverse processes of the higher lying lumbar and thoracic vertebrae

2 **Iliocostalis (muscles)**
- **Iliocostalis cervicis (muscle)** (2a): Projects from the cranial ribs to the transverse processes of the medial cervical vertebrae
- **Iliocostalis thoracis (muscle)** (2b): Projects from the lower ribs to the upper ribs
- **Iliocostalis lumborum (muscle)** (2c): Projects from the sacral bone and the iliac crest to the caudal ribs

6.2.10 Lateral Column – Angular System

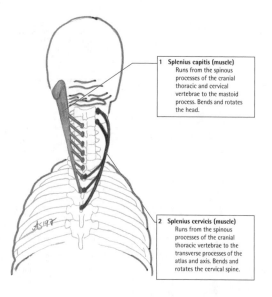

1 Splenius capitis (muscle)
Runs from the spinous processes of the cranial thoracic and cervical vertebrae to the mastoid process. Bends and rotates the head.

2 Splenius cervicis (muscle)
Runs from the spinous processes of the cranial thoracic vertebrae to the transverse processes of the atlas and axis. Bends and rotates the cervical spine.

6.2.11 Musculature of Back of Neck I

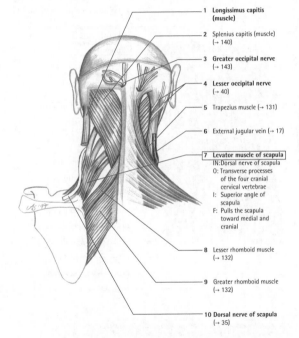

1 **Longissimus capitis (muscle)**

2 Splenius capitis (muscle) (→ 140)

3 **Greater occipital nerve** (→ 143)

4 **Lesser occipital nerve** (→ 40)

5 Trapezius muscle (→ 131)

6 External jugular vein (→ 17)

7 Levator muscle of scapula
IN: Dorsal nerve of scapula
O: Transverse processes of the four cranial cervical vertebrae
I: Superior angle of scapula
F: Pulls the scapula toward medial and cranial

8 Lesser rhomboid muscle (→ 132)

9 Greater rhomboid muscle (→ 132)

10 **Dorsal nerve of scapula** (→ 35)

6.2.12 Musculature of Back of Neck II

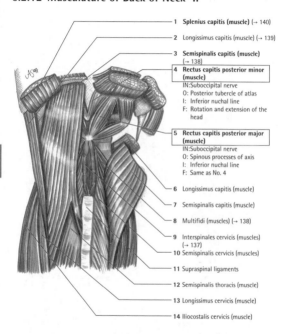

1 **Splenius capitis (muscle)** (→ 140)

2 Longissimus capitis (muscle) (→ 139)

3 **Semispinalis capitis (muscle)** (→ 138)

4 **Rectus capitis posterior minor (muscle)**
IN: Suboccipital nerve
O: Posterior tubercle of atlas
I: Inferior nuchal line
F: Rotation and extension of the head

5 **Rectus capitis posterior major (muscle)**
IN: Suboccipital nerve
O: Spinous processes of axis
I: Inferior nuchal line
F: Same as No. 4

6 Longissimus capitis (muscle)

7 Semispinalis capitis (muscle)

8 Multifidi (muscles) (→ 138)

9 Interspinales cervicis (muscles) (→ 137)

10 Semispinalis cervicis (muscles)

11 Supraspinal ligaments

12 Semispinalis thoracis (muscle)

13 Longissimus cervicis (muscle)

14 Iliocostalis cervicis (muscle)

The **short muscles of back of neck** consist of: Superior oblique muscle of head, inferior oblique muscle of head, rectus capitis posterior minor (muscle), rectus capitis posterior major (muscle) and rectus capitis lateralis (muscle).

6.2.13 Musculature of Back of Neck III / Nerves of Back of Neck

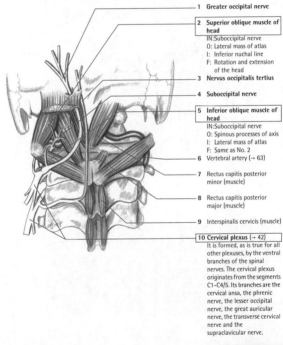

1 Greater occipital nerve

2 **Superior oblique muscle of head**
IN:Suboccipital nerve
O: Lateral mass of atlas
I: Inferior nuchal line
F: Rotation and extension of the head

3 Nervus occipitalis tertius

4 Suboccipital nerve

5 **Inferior oblique muscle of head**
IN:Suboccipital nerve
O: Spinous processes of axis
I: Lateral mass of atlas
F: Same as No. 2

6 Vertebral artery (→ 63)

7 Rectus capitis posterior minor (muscle)

8 Rectus capitis posterior major (muscle)

9 Interspinalis cervicis (muscle)

10 Cervical plexus (→ 42)
It is formed, as is true for all other plexuses, by the ventral branches of the spinal nerves. The cervical plexus originates from the segments C1-C4/5. Its branches are the cervical ansa, the phrenic nerve, the lesser occipital nerve, the great auricular nerve, the transverse cervical nerve and the supraclavicular nerve.

6.2.14 Vertebral Artery

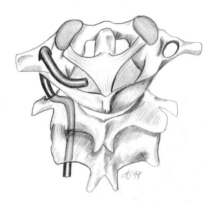

The vertebral artery (→ 143) is the first departing branch of the subclavian artery and projects, surrounded by the anterior scalenus muscles (→ 36) and longus colli muscle, in a cranial direction. It traverses the foramen of the transverse process at the 6th cervical vertebra and projects towards the atlas.

After leaving the transverse foramen at the most superior cervical vertebra, it makes a turn and projects in a medial direction along the vertebral groove of atlas.
It traverses the atlantooccipital membrane and then enters the subarachnoid space.
After entering the interior of the cranium through the great foramen, it joins with the vertebral artery from the opposite side, to form the basilar artery. Via the posterior cerebral arteries, the basilar artery is interconnected with the arterious circle of cerebrum (circle of Willis) (→ 63)

7. Thoracic Cavity and Viscera

7.1 Mediastinum

7.1.1 Structures of Mediastinum (Overview)

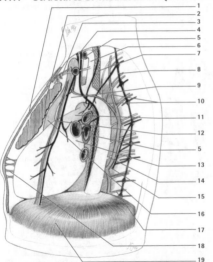

1	Sternum	11	Pulmonary trunk
2	Thymus	12	Pericardiacophrenic artery/vein arising from the left internal thoracic artery
3	Internal thoracic artery/vein (→ 32)		
4	Phrenic nerve (→ 42)	13	Heart (→ 158)
5	Vagus nerve (→ 148)	14	Posterior intercostal artery
6	Subclavian artery (→ 32)	15	Azygous vein (→ 204)
7	Esophagus (→ 175)	16	Superior vena cava (→ 204)
8	Sympathetic trunk	17	Posterior intercostal vein
9	Arch of aorta (→ 148)	18	Sternocardial ligaments
10	Accessory hemiazygous vein (→ 204)	19	Diaphragm (→ 155)

7.1.2 Contents and Subdivisions of Mediastinum

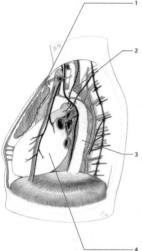

1 Anterior mediastinum
Between the pericardial sac and thoracic wall;
contains:
- Sternocardial ligaments
- Internal thoracic artery and vein

2 Superior mediastinum
Between the upper thoracic aperture and an
imaginary plane above the heart; contains:
- Vagus (→ 148) and phrenic nerve
- Sympathetic trunks
- Arch of aorta (→ 148) and branches
- Superior vena cava and brachiocephalic vein
- Trachea (→ 149)
- Esophagus (→ 175)
- Thoracic duct (→ 196)
- Thymus

3 Posterior mediastinum
Between the dorsal side of the heart and
the thoracic spine; contains:
- Vagal trunk (→ 148)
- Esophageal plexus of vagus nerve
- Sympathetic trunks
- Descending part of aorta (→ 148)
- Azygous and hemiazygous veins (→ 204)
- Esophagus (→ 175)
- Thoracic duct (→ 196)

4 Middle mediastinum
Between the dorsal side of the heart and
the thoracic spine; contains:
- Heart with pericardium (→ 161)
- Pericardiac branch of vagus nerve
- Phrenic nerves
- Pericardiac branch of phrenic nerve
- Pericardiacophrenic artery and vein

The **mediastinum** occupies the space between the right and left pleural cavity. It borders on the following structures:	
Ventral	Sternum
Dorsal	Thoracic spine
Lateral	Mediastinal pleura
Cranial	Superior thoracic aperture
Caudal	Diaphragm

7.1.3 Dorsal View of Mediastinum

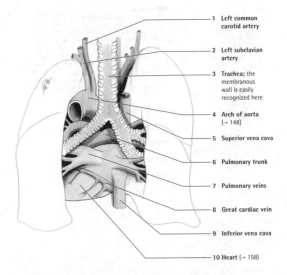

1 Left common carotid artery

2 Left subclavian artery

3 Trachea; the membranous wall is easily recognized here

4 Arch of aorta (→ 148)

5 Superior vena cava

6 Pulmonary trunk

7 Pulmonary veins

8 Great cardiac vein

9 Inferior vena cava

10 Heart (→ 158)

Detailed understanding of **mediastinum topography** is very important. A dorsal view of the mediastinum is shown in the figure above; for better visibility of the trachea the esophagus was removed. The descending aorta, also removed, would lie dorsal to the esophagus (→ 175). Therefore the order in which these structures are arranged is (from ventral to dorsal):

Ascending part of aorta – pulmonary trunk or right pulmonary artery – trachea

7.1.4 Vagus Nerve and Thoracic Part of Aorta

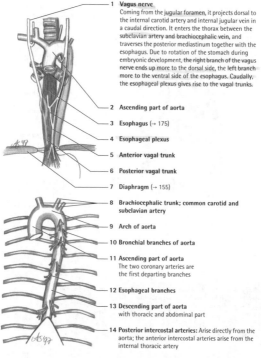

1 Vagus nerve
Coming from the jugular foramen, it projects dorsal to the internal carotid artery and internal jugular vein in a caudal direction. It enters the thorax between the subclavian artery and brachiocephalic vein, and traverses the posterior mediastinum together with the esophagus. Due to rotation of the stomach during embryonic development, the right branch of the vagus nerve ends up more to the dorsal side, the left branch more to the ventral side of the esophagus. Caudally, the esophageal plexus gives rise to the vagal trunks.

2 Ascending part of aorta

3 Esophagus (→ 175)

4 Esophageal plexus

5 Anterior vagal trunk

6 Posterior vagal trunk

7 Diaphragm (→ 155)

8 Brachiocephalic trunk; common carotid and subclavian artery

9 Arch of aorta

10 Bronchial branches of aorta

11 Ascending part of aorta
The two coronary arteries are the first departing branches.

12 Esophageal branches

13 Descending part of aorta
with thoracic and abdominal part

14 Posterior intercostal arteries: Arise directly from the aorta; the anterior intercostal arteries arise from the internal thoracic artery

See below for venous system of thorax and abdomen (→ 204)

7.2 Trachea, Lung, Diaphragm

7.2.1 Trachea

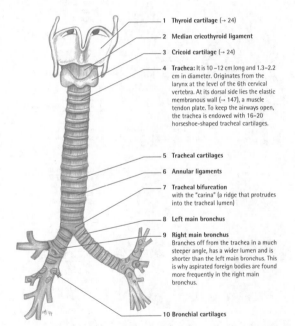

1 **Thyroid cartilage** (→ 24)

2 **Median cricothyroid ligament**

3 **Cricoid cartilage** (→ 24)

4 **Trachea:** It is 10–12 cm long and 1.3–2.2 cm in diameter. Originates from the larynx at the level of the 6th cervical vertebra. At its dorsal side lies the elastic membranous wall (→ 147), a muscle tendon plate. To keep the airways open, the trachea is endowed with 16–20 horseshoe-shaped tracheal cartilages.

5 **Tracheal cartilages**

6 **Annular ligaments**

7 **Tracheal bifurcation** with the "carina" (a ridge that protrudes into the tracheal lumen)

8 **Left main bronchus**

9 **Right main bronchus** Branches off from the trachea in a much steeper angle, has a wider lumen and is shorter than the left main bronchus. This is why aspirated foreign bodies are found more frequently in the right main bronchus.

10 **Bronchial cartilages**

7.2.2 Lung I

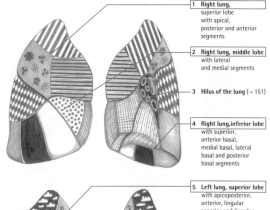

1 **Right lung,**
superior lobe
with apical,
posterior and anterior
segments

2 **Right lung, middle lobe**
with lateral
and medial segments

3 **Hilus of the lung** (→ 151)

4 **Right lung, inferior lobe**
with superior,
anterior basal,
medial basal, lateral
basal and posterior
basal segments

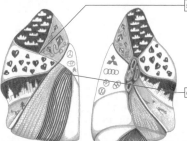

5 **Left lung, superior lobe**
with apicoposterior,
anterior, lingular
superior and lingular
inferior segments

6 **Left lung, inferior lobe**
with superior,
anterior basal, medial
basal, lateral
basal and posterior
basal segments

It is important to realize that the **right lung has three lobes**, the **left lung**
however only **two lobes** (the lingula is equivalent to the middle lobe of the right
lung)! As indicated in the figure, the apicoposterior segment of the left lung is
actually composed of two segments.

7.2.3 Lung II

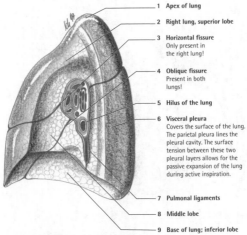

1 **Apex of lung**

2 **Right lung, superior lobe**

3 **Horizontal fissure**
Only present in
the right lung!

4 **Oblique fissure**
Present in both
lungs!

5 **Hilus of the lung**

6 **Visceral pleura**
Covers the surface of the lung.
The parietal pleura lines the
pleural cavity. The surface
tension between these two
pleural layers allows for the
passive expansion of the lung
during active inspiration.

7 **Pulmonal ligaments**

8 **Middle lobe**

9 **Base of lung; inferior lobe**

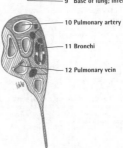

10 **Pulmonary artery**

11 **Bronchi**

12 **Pulmonary vein**

The **hilus of the lung** is
located on the mediastinal
lung surface. Within this
area the **bronchi, vessels
and nerves** enter or exit.
**The pulmonary veins lie
anterior and inferior, the
pulmonary arteries
medial and superior, and
the bronchi posterior in
the hilus.** The parietal
pleura and the visceral
pleura become continuous
with each other at the
pulmonary ligament.

7.2.4   Fine Structure of Lung

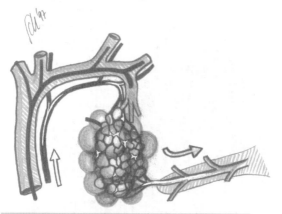

Two separate inflows supply the lung with blood:

- **The pulmonary artery** (→ 156) carries poorly oxygenated blood from the right heart to the lung. **Its branches follow the ramifications of the bronchial tree.**
- **The bronchial branches** of the descending aorta (→ 148) supply well-oxygenated blood for the nutrition of the airways (as far distal as the respiratory-bronchioles), the visceral pleura and the interlobular connective tissues. **These vessels run along the bronchi. Venous drainage of blood** from the lungs is performed by the pulmonary veins (→ 147) which carry well-oxygenated blood from the lung to the left atrium of the heart.

The **pulmonary veins** do not follow the bronchi, but run along the interlobular and intersegmental septa.
The ramifications of the bronchial tree from superior to inferior are as follows:

Trachea – main bronchi – lobar bronchi – segmental bronchi – subsegmental bronchi – small bronchi - (until here the walls of the airways are cartilage-reinforced, further below they are cartilage-free!) – bronchioles – terminal bronchioles – respiratory bronchioles – alveolar ducts – alveoli.

7.2.5 Boundaries of Pleurae and Lung

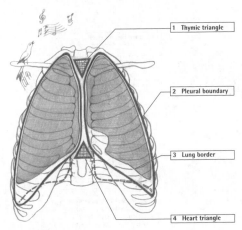

1 Thymic triangle

2 Pleural boundary

3 Lung border

4 Heart triangle

Position during expiration

	Sternal line	Medio-clavicular line	Middle axillary line	Scapular line	Paraverte-bral line
Lung border	4th rib (left) 6th rib (right)	6th rib	8th rib	10th rib	11th rib
Pleural boundary	6th rib	7th rib	9th rib	11th rib	12th rib

The two **triangles** in the above figure labeled 1 and 4 are **pleura devoid** areas that are not part of the pleural cavities. The table specifies **boundaries of the lung and pleura** as they project themselves onto the thorax.

This information is clinically important, when accessing the respiratory phase dependent changes of lung borders during percussion.

7.2.6 Recesses of the Lung

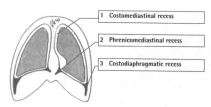

1	Costomediastinal recess
2	Phrenicomediastinal recess
3	Costodiaphragmatic recess

Expiration

| 1 | Costomediastinal recess |

Inspiration

Inspiration

Lung **recesses** are gaps between the parietal pleural lining of ribs, diaphragm and mediastinum. The two figures above show the recesses as they occur during **expiration**. Portions of the parietal pleura lie on top of each other and are separated by a narrow gap filled with liquid.

During **inspiration** the lung expands and protrudes into the recesses, pushing apart the stacked portions of the parietal pleura. These recesses therefore function as "spaces in reserve".

7.2.7 Caudal View of Diaphragm

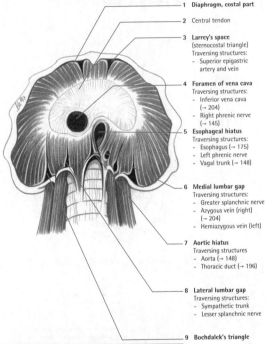

1 **Diaphragm, costal part**

2 **Central tendon**

3 **Larrey's space**
(sternocostal triangle)
Traversing structures:
- Superior epigastric artery and vein

4 **Foramen of vena cava**
Traversing structures:
- Inferior vena cava (→ 204)
- Right phrenic nerve (→ 145)

5 **Esophageal hiatus**
Traversing structures:
- Esophagus (→ 175)
- Left phrenic nerve
- Vagal trunk (→ 148)

6 **Medial lumbar gap**
Traversing structures:
- Greater splanchnic nerve
- Azygous vein (right) (→ 204)
- Hemiazygous vein (left)

7 **Aortic hiatus**
Traversing structures:
- Aorta (→ 148)
- Thoracic duct (→ 196)

8 **Lateral lumbar gap**
Traversing structures:
- Sympathetic trunk
- Lesser splanchnic nerve

9 **Bochdalek's triangle**

The **diaphragm** is innervated by the phrenic nerve (C3–5). It arises from the sternum, ribs VII-XII and vertebral bodies, and inserts at the central tendon. The diaphragm is the chief **muscle of respiration. Bochdalek's triangle** is a structurally weak area within the muscle plate (increased likelihood of herniation!).

7.3 Cardiovascular Systems

7.3.1 The Circulatory System in Adults

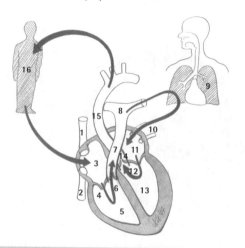

The abdominal organs are unusual, in that they first drain via the portal vein into the liver.

1	Superior vena cava	10	Pulmonary veins
2	Inferior vena cava	11	Left atrium
3	Right atrium	12	Left atrioventricular valve
4	Right atrioventricular valve	13	Left ventricle
5	Right ventricle	14	Aortic valve
6	Pulmonary valve	15	Aorta
7	Pulmonary trunk	16	Arteries – arterioles – capillaries of body periphery – venules – veins
8	Pulmonary arteries		
9	Lungs		

This ensures that substances within the venous blood are transported to the liver for further processing. The liver has its own independent arterial supply. Below, the cardiovascular system of the human adult will be discussed in more detail.

7.3.2 Fetal Circulation

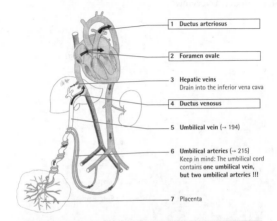

1 **Ductus arteriosus**

2 **Foramen ovale**

3 **Hepatic veins**
Drain into the inferior vena cava

4 **Ductus venosus**

5 Umbilical vein (→ 194)

6 Umbilical arteries (→ 215)
Keep in mind: The umbilical cord
contains **one umbilical vein,
but two umbilical arteries !!!**

7 Placenta

Due to the fact that the fetus receives oxygen not from its lungs, but from the mother via the placenta, **fetal circulation** exhibits the following special features:

1. Oxygenated blood originating from the placenta travels along the umbilical vein and **ductus venosus** (→ 180), and then empties, with most of the blood bypassing the liver, into the inferior vena cava, where it is mixed with venous unoxygenated blood from the body periphery. The fetal organism is adapted to these, by comparison to the adult, lower levels of oxygen (c. g. Hb F).
2. Some blood coming from the right atrium crosses the foramen ovale, located within the interatrial septum, and enters directly into the left atrium bypassing the lungs.
3. Blood from the pulmonary trunk is directed through the **ductus arteriosus** (→ 158) where most blood flow then enters the aorta, once again bypassing the lungs.

These bypasses are closed postnatally. The foramen ovale closes due to changes in circulatory pressure, after initiation of pulmonary respiration. The ductus venosus obliterates and the ductus arteriosus is closed actively at first (contraction of smooth muscle cells), but later obliterates too. Frequently a persistent ductus arteriosus is observed, which then is closed by surgery if necessary.

7.4   Heart, Pericardium

7.4.1   Heart, Overview

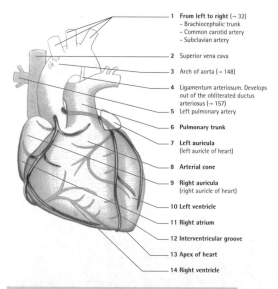

1 **From left to right** (→ 32)
   - Brachiocephalic trunk
   - Common carotid artery
   - Subclavian artery

2 Superior vena cava

3 Arch of aorta (→ 148)

4 Ligamentum arteriosum. Develops out of the obliterated ductus arteriosus (→ 157)

5 Left pulmonary artery

6 **Pulmonary trunk**

7 **Left auricula**
   (left auricle of heart)

8 **Arterial cone**

9 **Right auricula**
   (right auricle of heart)

10 **Left ventricle**

11 **Right atrium**

12 **Interventricular groove**

13 **Apex of heart**

14 **Right ventricle**

The **heart** is located behind the sternum in the middle mediastinum. Approximately 1/3 of the organ lies in the right and approx. 2/3 lies in the left half of the thorax. The heart axis roughly forms a 40° angle with respect to the median and frontal plane. Therefore the left atrium is not visible from a ventral position. The heart is about the size of an individual person's fist and weighs, on average, 350g. The heart is covered by the **epicardium**. A fold at the top base of the heart marks the zone, where the epicardium becomes continuous with the **pericardium** (heart sac). The previously discussed coronary vessels (→ 159) are embedded within the subepicardial adipose tissue.

The above-mentioned auricles are outpouchings of the atria.

7.4.2 Coronary Arteries

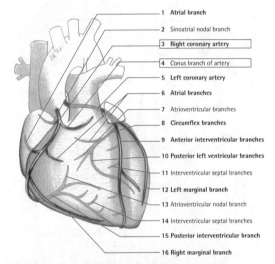

1 Atrial branch

2 Sinoatrial nodal branch

3 Right coronary artery

4 Conus branch of artery

5 Left coronary artery

6 Atrial branches

7 Atrioventricular branches

8 Circumflex branches

9 Anterior interventricular branches

10 Posterior left ventricular branches

11 Interventricular septal branches

12 Left marginal branch

13 Atrioventricular nodal branch

14 Interventricular septal branches

15 Posterior interventricular branch

16 Right marginal branch

The right and left **coronary arteries** arise cranial to the semilunar valve from the corresponding right and left aortic sinuses (→ 164). Generally each coronary artery supplies both the atrial and ventricular wall of its side and a part of the septum. However, frequently the right **coronary artery (RCA)** also supplies a section of the left rear ventricular wall and the **left coronary artery (LCA)** also supplies a section of the right front ventricular wall. The main stations of the cardiac conducting system are most frequently supplied by the right coronary artery. Please note: The **"LCA/RCA equal supply type"** described here, represents approximately 70% of all cases. However, in 20% of cases a **"LCA supply type "** (the posterior interventricular branch arises from the left coronary artery) and in 10% of cases a **"RCA supply type"** (the rear wall of the left ventricle is primarily supplied by the right coronary artery) is observed.

7.4.3 Veins of Heart (Dorsal View) I

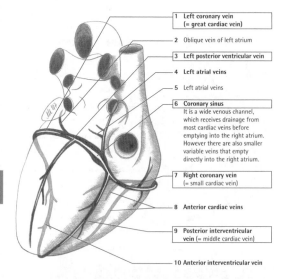

1 **Left coronary vein
(= great cardiac vein)**

2 Oblique vein of left atrium

3 **Left posterior ventricular vein**

4 **Left atrial veins**

5 Left atrial veins

6 **Coronary sinus**
It is a wide venous channel, which receives drainage from most cardiac veins before emptying into the right atrium. However there are also smaller variable veins that empty directly into the right atrium.

7 **Right coronary vein
(= small cardiac vein)**

8 Anterior cardiac veins

9 **Posterior interventricular vein (= middle cardiac vein)**

10 Anterior interventricular vein

As mentioned before, the coronary arteries and cardiac veins lie within the subepicardial adipose tissue. Due to anatomic variation, the coronary arteries may also run intramurally.

Clinical Information

It is remarkable, that for unknown reasons, intramural passages are not affected by the arteriosclerotic process. Infarction results, when a coronary artery becomes completely occluded. Besides drug therapy, balloon dilatation catheterization and bypass surgery are frequently used to prevent occlusal.

7.4.4 The Pericardial Sinus

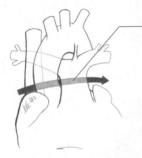

1 Transverse pericardial sinus
The pericardium (or heart sac) consists of the fibrous pericardium, serous pericardium and parietal layer. It becomes continuous with the epicardium at a fold, located on the walls of the great vessels above the base of the heart. The transverse sinus is a passage within the pericardial cavity, which is located between the aorta and pulmonary trunk on one side and the superior vena cava and pulmonary veins on the other side. The transitional folds located on these vessels form an enclosure for the transverse sinus.

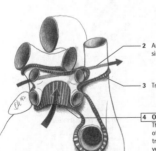

2 Arrow within the transverse pericardial sinus

3 Transitional fold of the pericardium

4 Oblique pericardial sinus
This sinus is formed by the dorsal wall of the pericardium and by the transitional fold of the pulmonary veins (usually 4).

7.4.5 Internal Chambers of Heart I

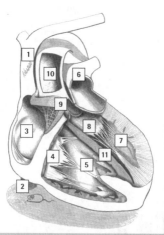

This figure provides an overview of the internal chambers of the heart. The structures labeled with numbers are itemized below:

1	Superior vena cava (→ 204)	7	Left ventricle
2	Inferior vena cava (→ 204)	8	Left atrioventricular valve = mitral valve
3	Right atrium	9	Aortic valve
4	Right atrioventricular valve = tricuspid valve	10	Ascending aorta (→ 148)
5	Right ventricle	11	Interventricular septum
6	Pulmonary trunk		

7.4.6 Internal Chambers of Heart II

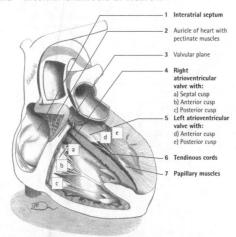

1 **Interatrial septum**

2 **Auricle of heart with pectinate muscles**

3 **Valvular plane**

4 **Right atrioventricular valve with:**
a) Septal cusp
b) Anterior cusp
c) Posterior cusp

5 **Left atrioventricular valve with:**
d) Anterior cusp
e) Posterior cusp

6 **Tendinous cords**

7 **Papillary muscles**

The tendinous cords project to the papillary muscles. The anterior cusp for example inserts into the anterior and posterior papillary muscle.

The internal chambers of the heart are lined by the endocardium. Even the tendinous cords (thread-like bands of fibrous tissue coming from the papillary muscles) are covered by this epithelial layer. The tendinous cords and the papillary muscles prevent the prolapsing of the atrioventricular valves into the atria during systole (ventricular emptying).

The walls of the **atria** are **smooth** except for the atrial auricles, which have a defined wall structure formed by the pectinate muscles. The walls of the **ventricles** however have a **pronounced relief**, due to the presence of the **trabeculae carneae** (small fleshy ridges of heart muscle) and the papillary muscles. As demonstrated in the figure above, the myocardium of the left ventricle is, with a wall thickness of approximately 15 mm, roughly three times as thick as the one of the right ventricle. This is due to the much higher pressure the left ventricle must generate. The right atrium receives drainage from the upper and lower vena cava and the coronary sinus. In adults the interatrial septum has a groove, the oval fossa, which is a remnant of the fetal foramen ovale.

7.4.7 Valvular Plane of Heart

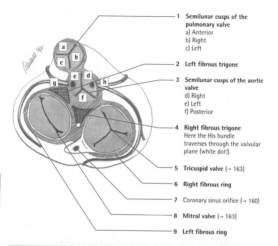

1 Semilunar cusps of the
  pulmonary valve
  a) Anterior
  b) Right
  c) Left

2 Left fibrous trigone

3 Semilunar cusps of the aortic
  valve
  d) Right
  e) Left
  f) Posterior

4 Right fibrous trigone
  Here the His bundle
  traverses through the valvular
  plane (white dot!)

5 Tricuspid valve (→ 163)

6 Right fibrous ring

7 Coronary sinus orifice (→ 160)

8 Mitral valve (→ 163)

9 Left fibrous ring

All four cardiac valves lie in one plane. This so-called **valvular plane** (or **skeleton of heart**) projects onto a line, which extends from the right sternal attachment of the 6th rib to the left sternal attachment of the 3rd rib. It constitutes an **electrically insulating layer** between the atria and ventricles, which is only interrupted by the traversing His bundle at the right fibrous trigone. The fibrous rings surround the atrioventricular orifices and give rise to the atrioventricular valves. As shown in the figure above, both coronary arteries (g and h) arise directly above the aortic valve out of the aortic sinus.

The **eustachian valve** is a fold within the right atrium which directs blood from the inferior vena cava into and through the foramen ovale, during the fetal period. It is only mentioned here for completeness.

Clinical Information

In rare cases, the electrically insulating layer is interrupted by additional bundles (bundle of Kent and James). This in turn causes arrhythmias (e.g. Wolff-Parkinson-White syndrome = WPW-syndrome)

7.4.8 Cardiac Valves

The mitral valve

The left atrioventricular valve's alternative name is derived from the valve's resemblance to a catholic bishop's miter (headdress).
It separates the left atrium from the left ventricle and has two cusps, **anterior** and **posterior**, which is why it is also called **bicuspid valve.**

The tricuspid valve

This atrioventricular valve, as the name indicates, has three cusps, **anterior, posterior** and **septal**. It separates the right atrium from the right ventricle. Again it should be made clear here, that **each cusp sends off tendinous cords to two papillary muscles**. This is not shown correctly in the figure for reasons of clarity.

The aortic and pulmonary valve

These valves are formed by a **duplication of the endocardium**. They consist of **three semilunar cusps** and lie within the path of ventricular outflow at the beginning of the aorta and pulmonary artery. They prevent blood from flowing back into the chambers at the end of systole. The mechanism responsible for this process is indicated in the figure by an arrow. The aortic sinus lies above the aortic valve and gives rise to the **coronary vessels.**

7.4.9 Pacemaker and Conducting System

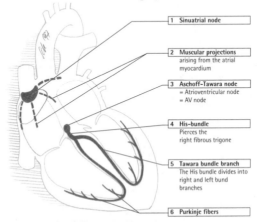

1 Sinuatrial node	
2 Muscular projections arising from the atrial myocardium	
3 Aschoff-Tawara node = Atrioventricular node = AV node	
4 His-bundle Pierces the right fibrous trigone	
5 Tawara bundle branch The His bundle divides into right and left bund branches	
6 Purkinje fibers	

The pacemaker and conducting system of the heart does **not** consist of nerve fibers, but of specialized cardiac muscle cells. The **sinuatrial (SA) node** is located ventral to the opening of the superior vena cava. The SA node is the pacemaker of the heart and gives off an impulse 60–90 times per minute. The **atrioventricular (AV) node** lies in the dorsal wall of the right atrium. Its outstanding feature is that it delays the transmission of arriving impulses, thereby allowing coordinated atrial and ventricular actions. The **His bundle** pierces the electrically insulating layer of the heart skeleton, transitions into the **Tawara bundle branch** and ends in the **purkinje fibers**, a terminal network of fibers, which propagate the excitation over the ventricular myocardium from heart apex towards heart base. Parts of the conducting systems, downstream of the sinuatrial node, are also able to initiate impulses albeit with lower intrinsic frequency (AV-node: 40 per minute; His-bundle: 20 per minute). Therefore under normal conditions, these centers are always triggered into action by the more rapid build-up of excitation in the sinuatrial node, before they have a chance to trigger themselves. If for any reason the sinuatrial node fails, the AV node substitutes as a secondary pacemaker albeit with lower frequency. The cardiac conducting system is sensitive to the **actions of the vegetative nervous systems.**

7.4.10 Cardiac Auscultation Points

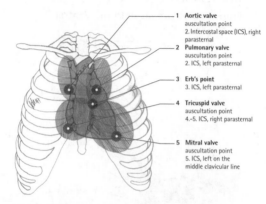

1 Aortic valve
auscultation point
2. Intercostal space (ICS), right parasternal

2 Pulmonary valve
auscultation point
2. ICS, left parasternal

3 Erb's point
3. ICS, left parasternal

4 Tricuspid valve
auscultation point
4.-5. ICS, right parasternal

5 Mitral valve
auscultation point
5. ICS, left on the
middle clavicular line

The heart sounds and murmurs are caused by the action of the heart valves and by the blood flow. Normally occurring sound phenomena, caused by movement of the valves, are called **heart sounds,** while all other remaining sound phenomena (but not only pathologic ones) are called **heart murmurs.**

The sounds produced by a cardiac valve are best heard at the auscultation points.

At Erb's point the sounds produced by all four cardiac valves can be heard.
At this point auscultation allows to trace back to the cardiac valves contributing to an occurring murmur

The **cardiac auscultation points** are **not** congruent with the cardiac valve's projection onto the thoracic wall (see the heart silhouette in the above figure)!
Due to patient variability they are not clearly defined points, but as indicated here, areas along the thoracic wall. Their approximate locations are indicated above.

8.   Abdominal Cavity and Viscera

8.1   General Facts

8.1.1   Projection of Organs onto the Ventral Trunk Wall

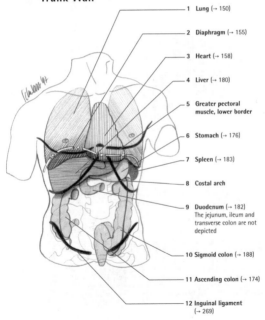

1 Lung (→ 150)

2 Diaphragm (→ 155)

3 Heart (→ 158)

4 Liver (→ 180)

5 Greater pectoral muscle, lower border

6 Stomach (→ 176)

7 Spleen (→ 183)

8 Costal arch

9 Duodenum (→ 182)
   The jejunum, ileum and transverse colon are not depicted

10 Sigmoid colon (→ 188)

11 Ascending colon (→ 174)

12 Inguinal ligament (→ 269)

8.1.2 Overview of Abdominal Cavity

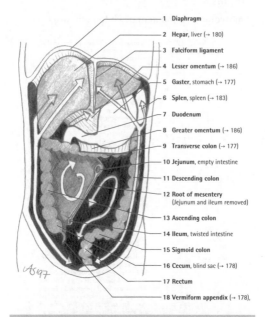

1 Diaphragm
2 Hepar, liver (→ 180)
3 Falciform ligament
4 Lesser omentum (→ 186)
5 Gaster, stomach (→ 177)
6 Splen, spleen (→ 183)
7 Duodenum
8 Greater omentum (→ 186)
9 Transverse colon (→ 177)
10 Jejunum, empty intestine
11 Descending colon
12 Root of mesentery
    (Jejunum and ileum removed)
13 Ascending colon
14 Ileum, twisted intestine
15 Sigmoid colon
16 Cecum, blind sac (→ 178)
17 Rectum
18 Vermiform appendix (→ 178),

The **peritoneal cavity**, which is lined by the **peritoneum**, is divided into the upper and lower abdominal region. The **upper abdominal region** extends from the 9th thoracic to the 2nd lumbar vertebra. The **lower abdominal** region continues from the 2nd lumbar vertebra to the beginning of the pelvic area. The **root** of the transverse mesocolon marks the border between these two cavities.

Remark: The root of mesentery is only about 20 cm long, while the duodenum has a length of 4–5 m.

8.2 Embryonic Development of Abdominal Organs

8.2.1 Embryonic Development of Abdominal Organs I

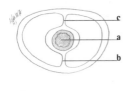

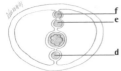

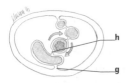

To better understand the positions of the organs within the abdominal cavity, the organization of the peritoneum as well as the course of the mesenteries and other structures, we will briefly recapitulate the embryonic development of this area.

The **intestinal tube** (a) is formed out of the yolk sac by ventral infolding. It continues to communicate with the yolk sac through the **omphaloenteric duct**, which in some adults persists as **Meckel's diverticula**. The intestinal tube is suspended ventrally (b) and dorsally (c) by mesenteries. The caudal part of the ventral mesentery is eliminated during development (the cranial part persists) and **only** one undivided abdominal cavity remains. The **ventral mesentery** gives rise to the **liver** (d). The **spleen** (e) and pancreas (f) develop out of the **dorsal mesentery**. The mesos give rise to the peritoneal lining of the organs.

The intestinal tube extends, relative to the **sagittal plane**, in length and width, to form the subdivisions, which will later contribute to the distinct portions of the intestine. The stomach, lying in the **medial plane**, now begins to rotate a total of 90°. The liver is shifted to the right. The ventral meso arises from the ventral side of the liver as **falciform ligament** (g) (→ 180), from the dorsal side as **lesser omentum** (h) (→ 186). The dorsal meso as well as the descendant spleen and pancreas are shifted to the left. The lesser omentum combines with the dorsal meso to form the **omental bursa** (i). The parietal peritoneum fuses with the peritoneal lining of the pancreas and the pancreas becomes secondarily retroperitoneal (j).

8.2.2 Embryonic Development of Abdominal Organs II

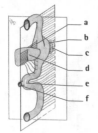

The **intestinal tube** (embryo approx. 5 mm crown-rump length (CRL), approx. at the end of the first month) is shown in the figure above. The enlargements occurring within the sagittal plane have already established the **lesser** (a) and **greater** (b) **curvature** of the stomach (shown here **before** rotation!). The **duodenal loop** (c), a loop of the intestine oriented towards the navel with its apex bordering on the omphaloenteric ductus (f), can also be seen in the figure. Ingrowth of liver (d), spleen and pancreas (c) into the mesos has already occurred.

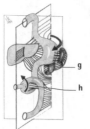

The figure in the middle (approx. 30 mm CRL, second month) shows the stomach **rotating 90°leftward** with the greater curvature initially pointing backwards. It now resides in a rather frontal than sagittal plane. The dorsal mesogastrium expands, and together with the lesser omentum forms the **omental bursa** (g). The ingrown dorsal meso subdivides into **gastrosplenic** and **phrenicosplenic ligaments**. Parallel to all these occurrences, the **intestine begins to rotate**, with the **superior mesenteric artery** acting as axis of rotation. Coming from dorsal, this artery runs within the mesentery of the duodenal or umbilical loop (h).

The figure below (CRL >50 mm, end of the third month) shows the **rotation of the intestine** in progress. It **rotates approximately 270° counter-clockwise**. Meanwhile, depending on the position along the intestinal tube, the various parts of the intestine extend differentially in length and width. The **ascending** (j) and **descending colon** (k) attach to the dorsal trunk wall and therefore become **secondarily retroperitoneal** (as is the case for the pancreas). Contrary to this, the **transverse colon** (i) remains **intraperitoneal** and later on comes to lie ventral to the stomach.

8.2.3 Embryonic Development of Abdominal Organs III

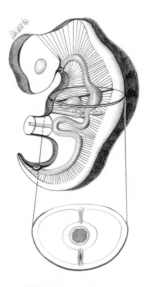

This figure and the figure on the following page further illustrate the above-mentioned events. Indicated in the figures are the levels of the transverse sections shown below. Keep in mind that the caudal part of the ventral meso regresses during development.

Also visible are the ventral and dorsal primordia of the pancreas, which later fuse with each other. For reasons of simplification, the ventral primordium was not mentioned until now.

8.2.4 Embryonic Development of Abdominal Organs IV

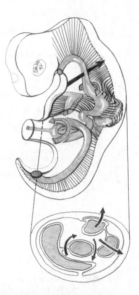

The arrows shown in the transverse section indicate the direction of rotation of the individual organs. In the above figure the black arrow located at the future site of the esophagus, indicates the sagittal plane. The hatched arrow in the figure indicates the direction of rotation.

8.3 Digestive Tract

8.3.1 Schematic of Digestive Tract

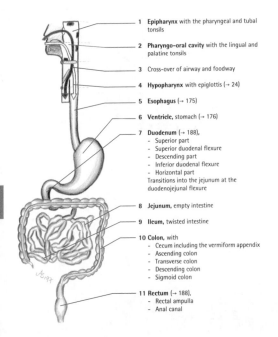

1 **Epipharynx** with the pharyngeal and tubal tonsils

2 **Pharyngo-oral cavity** with the lingual and palatine tonsils

3 **Cross-over of airway and foodway**

4 **Hypopharynx** with epiglottis (→ 24)

5 **Esophagus** (→ 175)

6 **Ventricle,** stomach (→ 176)

7 **Duodenum** (→ 188),
- Superior part
- Superior duodenal flexure
- Descending part
- Inferior duodenal flexure
- Horizontal part
Transitions into the jejunum at the duodenojejunal flexure

8 **Jejunum,** empty intestine

9 **Ileum,** twisted intestine

10 **Colon,** with
- Cecum including the vermiform appendix
- Ascending colon
- Transverse colon
- Descending colon
- Sigmoid colon

11 **Rectum** (→ 188),
- Rectal ampulla
- Anal canal

8.3.2 Dorsal View of Esophagus

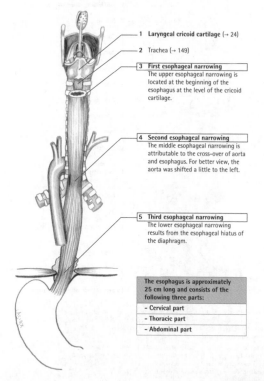

1 Laryngeal cricoid cartilage (→ 24)

2 Trachea (→ 149)

3 First esophageal narrowing
The upper esophageal narrowing is located at the beginning of the esophagus at the level of the cricoid cartilage.

4 Second esophageal narrowing
The middle esophageal narrowing is attributable to the cross-over of aorta and esophagus. For better view, the aorta was shifted a little to the left.

5 Third esophageal narrowing
The lower esophageal narrowing results from the esophageal hiatus of the diaphragm.

The esophagus is approximately 25 cm long and consists of the following three parts:

– Cervical part

– Thoracic part

– Abdominal part

8.3.3 Stomach

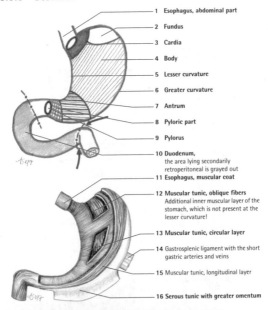

1 Esophagus, abdominal part
2 Fundus
3 Cardia
4 Body
5 Lesser curvature
6 Greater curvature
7 Antrum
8 Pyloric part
9 Pylorus
10 Duodenum,
the area lying secondarily
retroperitoneal is grayed out
11 Esophagus, muscular coat
12 Muscular tunic, oblique fibers
Additional inner muscular layer of the
stomach, which is not present at the
lesser curvature!
13 Muscular tunic, circular layer
14 Gastrosplenic ligament with the short
gastric arteries and veins
15 Muscular tunic, longitudinal layer
16 Serous tunic with greater omentum

The **gastric canal** consists of gastric folds that run along the lesser curvature.
The gastric mucosa contains three types of cells: The **chief cells** (produce pepsin),
the **parietal cells** (produce intrinsic factor and hydrochloric acid) and the mucous
neck cells (produce mucus).

Keep in mind: The following **wall layout** is found throughout the entire
gastrointestinal tract (from interior to exterior): **Mucous tunic – submucous
layer – inner circular layer of muscular tunic – outer longitudinal layer of
muscular tunic – subserous layer – serous tunic.**

8.3.4 Stomach and Colon

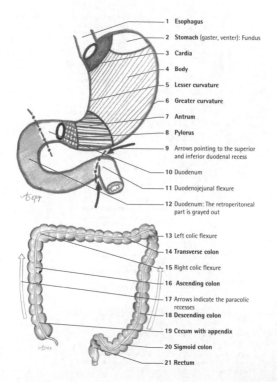

1 Esophagus
2 Stomach (gaster, venter): Fundus
3 Cardia
4 Body
5 Lesser curvature
6 Greater curvature
7 Antrum
8 Pylorus
9 Arrows pointing to the superior and inferior duodenal recess
10 Duodenum
11 Duodenojejunal flexure
12 Duodenum: The retroperitoneal part is grayed out

13 Left colic flexure
14 Transverse colon
15 Right colic flexure
16 Ascending colon
17 Arrows indicate the paracolic recesses
18 Descending colon
19 Cecum with appendix
20 Sigmoid colon
21 Rectum

8.3.5 Characteristics of Colon and Cecum

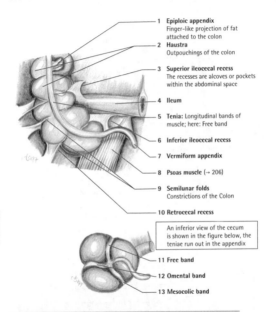

1 **Epiploic appendix**
Finger-like projection of fat attached to the colon
2 **Haustra**
Outpouchings of the colon
3 **Superior ileocecal recess**
The recesses are alcoves or pockets within the abdominal space
4 **Ileum**
5 **Tenia:** Longitudinal bands of muscle; here: Free band
6 **Inferior ileocecal recess**
7 **Vermiform appendix**
8 **Psoas muscle** (→ 206)
9 **Semilunar folds**
Constrictions of the Colon
10 **Retrocecal recess**

An inferior view of the cecum is shown in the figure below, the teniae run out in the appendix

11 **Free band**
12 **Omental band**
13 **Mesocolic band**

The following structures enable to distinguish macroscopically between colon and duodenum: Teniae, epiploic appendix, semilunar folds and haustra.

The semilunar folds and haustra result from peristaltic muscle contraction. The folds and haustra are therefore not static, but move along the colon!

The **Kerckring's folds** are folds of the mucosa which, beginning at the duodenum, steadily decrease in height towards distal and are absent in the colon. From the duodenum towards the colon, the **crypta** increase in depth.

8.3.6 Closure Mechanisms of Gastrointestinal Tract

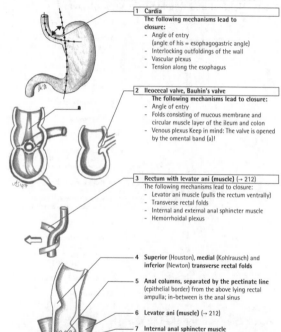

1 Cardia
The following mechanisms lead to closure:
- Angle of entry (angle of his = esophagogastric angle)
- Interlocking outfoldings of the wall
- Vascular plexus
- Tension along the esophagus

2 Ileocecal valve, Bauhin's valve
The following mechanisms lead to closure:
- Angle of entry
- Folds consisting of mucous membrane and circular muscle layer of the ileum and colon
- Venous plexus Keep in mind: The valve is opened by the omental band (a)!

3 Rectum with levator ani (muscle) (→ 212)
The following mechanisms lead to closure:
- Levator ani muscle (pulls the rectum ventrally)
- Transverse rectal folds
- Internal and external anal sphincter muscle
- Hemorrhoidal plexus

4 Superior (Houston), **medial** (Kohlrausch) and **inferior** (Newton) **transverse rectal folds**

5 Anal columns, separated by the pectinate line (epithelial border) from the above lying rectal ampulla; in-between is the anal sinus

6 Levator ani (muscle) (→ 212)

7 Internal anal sphincter muscle

8 External anal sphincter muscle (→ 213)

The closure mechanisms of the gastrointestinal tract function to regulate the transport of material and to prevent material from flowing back.

8.4   Liver, Gallbladder, Pancreas, Spleen

8.4.1   Ventral and Caudodorsal View of Liver

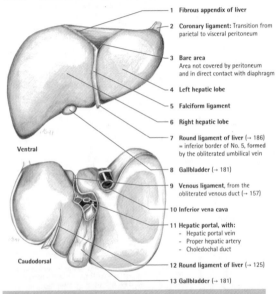

Ventral

Caudodorsal

1 **Fibrous appendix of liver**

2 **Coronary ligament:** Transition from parietal to visceral peritoneum

3 **Bare area**
Area not covered by peritoneum and in direct contact with diaphragm

4 **Left hepatic lobe**

5 **Falciform ligament**

6 **Right hepatic lobe**

7 **Round ligament of liver** (→ 186)
= inferior border of No. 5, formed by the obliterated umbilical vein

8 **Gallbladder** (→ 181)

9 **Venous ligament,** from the obliterated venous duct (→ 157)

10 **Inferior vena cava**

11 **Hepatic portal,** with:
- Hepatic portal vein
- Proper hepatic artery
- Choledochal duct

12 **Round ligament of liver** (→ 125)

13 **Gallbladder** (→ 181)

Mnemonic:

The structures 9–13 form an **"H"** shape, with No. 11 forming the crossbar of the "H".

The liver is the largest parenchymal organ and is located in the **right hypochondriac and epigastric region** of the right upper abdomen. The **bare area** constitutes a part of the diaphragmatic surface, which is not covered by the peritoneum, but instead directly adheres to the diaphragm. The **falciform ligament** is attached to the ventral trunkwall. In healthy individuals the inferior border of the liver coincides with the costal arch.

The liver receives blood via the **proper hepatic artery** and the **portal vein.** Venous drainage occurs via the **hepatic veins** into the **inferior vena cava.**

8.4.2 Biliary Ducts

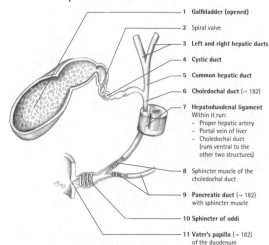

1 Gallbladder (opened)

2 Spiral valve

3 **Left and right hepatic ducts**

4 **Cystic duct**

5 **Common hepatic duct**

6 **Choledochal duct** (→ 182)

7 **Hepatoduodenal ligament**
Within it run:
- Proper hepatic artery
- Portal vein of liver
- Choledochal duct
  (runs ventral to the
  other two structures)

8 **Sphincter muscle of the
choledochal duct**

9 **Pancreatic duct** (→ 182)
with sphincter muscle

10 **Sphincter of oddi**

11 **Vater's papilla** (→ 182)
of the duodenum

Bile is produced in the liver cells and concentrated and stored in the gallbladder.
Release of bile into the duodenum is achieved through contraction of the gallbladder
muscles. The stimulus of contraction is **cholecystokinin**, released from the duodenum
on the entry of e.g. fat-containing chyme.

Clinical Information

Gallstones or tumors occurring within the area of the **major duodenal papilla**
(= **Vater's papilla**) and biliary ducts may impair bile outflow. This may cause bile to
accumulate back into the liver, eventually leading to posthepatic icterus.

The dilated common final pathway of the choledochal and pancreatic duct is called
hepatopancreatic ampulla. It is not consistently developed.

The **minor duodenal papilla** is the variable site at which the accessory pancreatic duct
drains into the duodenum (→ 182).

8.4.3 Pancreas

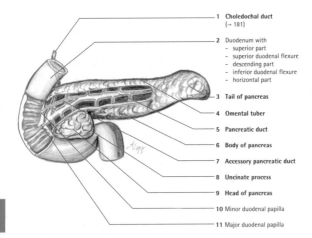

1 Choledochal duct
  (→ 181)

2 Duodenum with
  - superior part
  - superior duodenal flexure
  - descending part
  - inferior duodenal flexure
  - horizontal part

3 **Tail of pancreas**

4 **Omental tuber**

5 **Pancreatic duct**

6 **Body of pancreas**

7 **Accessory pancreatic duct**

8 **Uncinate process**

9 **Head of pancreas**

10 Minor duodenal papilla

11 Major duodenal papilla

The **pancreas** is located secondarily retroperitoneal. The head of the pancreas clings to the duodenal "C" and projects out as hook-shaped uncinate process.

The organ is comprised of an exo- and endocrine part: The **exocrine part** is purely serous and produces digestive enzyme. The Langerhans islets of the **endocrine part** secrete glucagon (alpha cells), insulin (beta cells) and somatostatin (delta cells).

Blood is supplied to the pancreas from the superior and inferior pancreaticoduodenal arteries (→ 189) and from the branches of the splenic artery (→ 183), which run along the upper border of the pancreatic body. The superior mesenteric artery and vein (→ 190) run between the lower pancreatic border and the upper border of the last duodenal segment.

The accessory pancreatic duct is formed by incomplete fusion of the ventral and dorsal pancreatic primordia (→ 170). Usually only one excretory duct, the pancreatic duct, develops.

8.4.4 Spleen

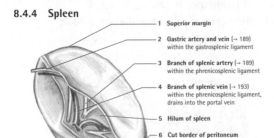

1 Superior margin

2 Gastric artery and vein (→ 189)
within the gastrosplenic ligament

3 Branch of splenic artery (→ 189)
within the phrenicosplenic ligament

4 Branch of splenic vein (→ 193)
within the phrenicosplenic ligament,
drains into the portal vein

5 Hilum of spleen

6 Cut border of peritoneum

7 Inferior margin

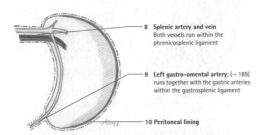

8 Splenic artery and vein
Both vessels run within the
phrenicosplenic ligament

9 Left gastro-omental artery; (→ 189)
runs together with the gastric arteries
within the gastrosplenic ligament

10 Peritoneal lining

The **spleen,** which arises from the dorsal meso of the stomach during embryogenesis,
is located **intraperitoneal.**

The spleen extends **parallel to the 10th rib.** It comprises one quarter of the total mass of
lymphatic tissue. It is the only organ within the human body in which erythrocytes enter
directly from the blood stream into the tissue. It singles out senescent red blood cells, degrades
them, and forwards the degradation products via the portal vein to the liver.

8.5   Peritoneum, Omenta

8.5.1   Arrangement of Peritoneum in the Adult

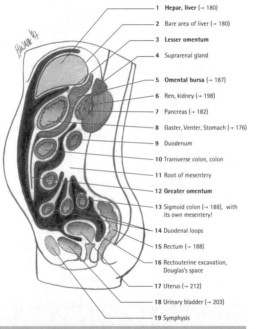

1 Hepar, liver (→ 180)

2 Bare area of liver (→ 180)

3 **Lesser omentum**

4 Suprarenal gland

5 **Omental bursa** (→ 187)

6 Ren, kidney (→ 198)

7 Pancreas (→ 182)

8 Gaster, Venter, Stomach (→ 176)

9 Duodenum

10 Transverse colon, colon

11 Root of mesentery

12 **Greater omentum**

13 Sigmoid colon (→ 188), with its own mesentery!

14 Duodenal loops

15 Rectum (→ 188)

16 Rectouterine excavation, Douglas's space

17 Uterus (→ 212)

18 Urinary bladder (→ 203)

19 Symphysis

The area marked in black in the above figure is the peritoneal cavity.

Easily recognizable is the formation of the greater omentum by fusion of two adjacent peritoneal layers. The pancreas as well as the kidney, duodenum and rectum are located retroperitoneal (→ 187). The **epiploic foramen** connects the peritoneal cavity with the omental bursa. In females the lowest point of the abdominal cavity is called **Douglas's space**. In males this point is called **rectovesical excavation**.

8.5.2   Arrangement of the Peritoneum

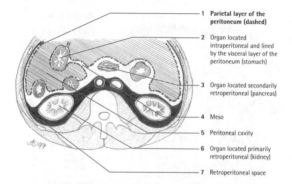

1 **Parietal layer of the peritoneum (dashed)**

2 Organ located intraperitoneal and lined by the visceral layer of the peritoneum (stomach)

3 Organ located secondarily retroperitoneal (pancreas)

4 Meso

5 Peritoneal cavity

6 Organ located primarily retroperitoneal (kidney)

7 Retroperitoneal space

Relationship of the abdominal organs to the peritoneum
Intraperitoneal organs
Stomach, parts of the duodenum, ileum, jejunum, vermiform appendix, cecum, transverse and sigmoid colon, liver, gallbladder and spleen
Secondarily retroperitoneal organs
Parts of the duodenum, pancreas, ascending colon, descending colon and rectum
Primarily retroperitoneal organs
Kidneys and suprarenal glands

Within the peritoneal cavity, the area between den organs is filled with **peritoneal fluid** (only a few milliliters!). It enables the intraperitoneal organs to move against each other essentially without friction. The direction of flow is indicated by arrows in the figure of the previous page (especially into the paracolic recess, the right and left mesentericocolic gap, into the omental bursa and into the subphrenic and subhepatic recess).

Clinical Information

Ascites is a condition in which an abnormal amount of fluid accumulates in the abdominal cavity. This can be caused by inflammation (e.g. peritonitis) or by other processes (e.g. heart insufficiency, cirrhosis of liver, trauma).

8.5.3 Lesser and Greater Omentum

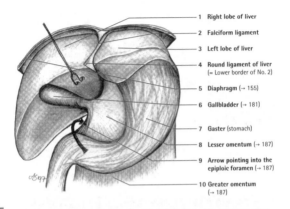

1 Right lobe of liver
2 Falciform ligament
3 Left lobe of liver
4 Round ligament of liver
  (= Lower border of No. 2)
5 Diaphragm (→ 155)
6 Gallbladder (→ 181)
7 Gaster (stomach)
8 Lesser omentum (→ 187)
9 Arrow pointing into the
  epiploic foramen (→ 187)
10 Greater omentum
  (→ 187)

The **lesser omentum** forms the anterior wall of the omental bursa. It lies almost exactly in the frontal plane. It projects from the liver to the lesser curvature of the stomach and to the beginning of the duodenum.
It consists of the following three parts:
- Hepatogastric ligament
- Hepatoduodenal ligament
- Gastrophrenic ligament (this most superior part of the lesser omentum runs between stomach and diaphragm)
The following structures run within the hepatoduodenal ligament:
- Proper hepatic artery
- Portal vein
- Choledochal duct
The **omental bursa** is the biggest recess of the abdominal cavity. It is a "sliding gap" that permits free movement of the stomach. The omental bursa is frequently used as an access path for surgery. Therefore exact knowledge of this areas configuration is very important.
The **greater omentum** consists of a double-layered sheet of peritoneum which, as can be seen clearly in the previous figure, is attached to the transverse colon and greater curvature of the stomach. In some areas the two peritoneal layers coalesce. The greater omentum extends to a variable degree into the lower abdomen.

8.5.4 Omental Bursa and its Boundaries

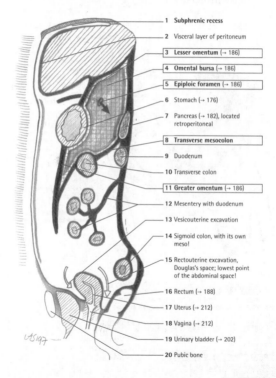

1 Subphrenic recess

2 Visceral layer of peritoneum

3 Lesser omentum (→ 186)

4 Omental bursa (→ 186)

5 Epiploic foramen (→ 186)

6 Stomach (→ 176)

7 Pancreas (→ 182), located retroperitoneal

8 Transverse mesocolon

9 Duodenum

10 Transverse colon

11 Greater omentum (→ 186)

12 Mesentery with duodenum

13 Vesicouterine excavation

14 Sigmoid colon, with its own meso!

15 Rectouterine excavation, Douglas's space; lowest point of the abdominal space!

16 Rectum (→ 188)

17 Uterus (→ 212)

18 Vagina (→ 212)

19 Urinary bladder (→ 202)

20 Pubic bone

8.5.5 Relationship of Sigmoid Colon and Rectum to the Peritoneum

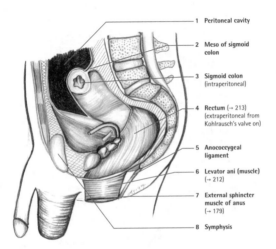

1 Peritoneal cavity

2 Meso of sigmoid colon

3 Sigmoid colon (intraperitoneal)

4 Rectum (→ 213) (extraperitoneal from Kohlrausch's valve on)

5 Anococcygeal ligament

6 Levator ani (muscle) (→ 212)

7 External sphincter muscle of anus (→ 179)

8 Symphysis

Intraperitoneal organs:

Stomach, parts of the duodenum, ileum, jejunum, vermiform appendix, cecum, transverse and sigmoid colon, liver, gallbladder, spleen and ovary

Secondarily retroperitoneal organs:

Parts of the duodenum, pancreas, ascending colon, descending colon and rectum

Primarily retroperitoneal organs:

Kidneys and suprarenal glands

8.6   Vessels, Nerves, Lymphatic System

8.6.1   Celiac Trunk

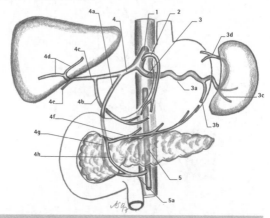

The **abdominal aorta** gives off the following unpaired visceral branches: The **celiac trunk**, the superior mesenteric artery (→ 190) and the **inferior mesenteric artery** (→ 191).

The **paired visceral branches** are: The inferior phrenic artery, the middle suprarenal artery (→ 200), the renal artery (→ 200) and the testicular or ovarian artery (→ 200).

The **celiac trunk** (1) gives off the following branches (2–4):			
2	**Left gastric artery**	4c	Proper hepatic artery
3	**Splenic artery** (meandering course) (→ 183)	4d	Right and left branches of 4c
3a	Pancreatic arteries (not depicted)	4e	Cystic artery
3b	Left gastroepiploic artery (= left gastroomental) (→ 183)	4f	Right gastroepiploic artery (= Right gastroomental)
3c	Arteries of the spleen (→ 183)	4g,h	Superior pancreaticoduodenal arteries
3d	Short anterior and posterior gastric arteries	5	Superior mesenteric artery (→ 190)
4	**Common hepatic artery**	5a	Inferior pancreaticoduodenal artery
4a	Gastroduodenal artery		
4b	Right gastric artery		

8.6.2 Superior Mesenteric Artery

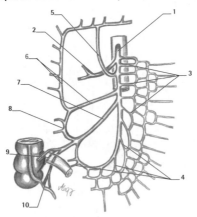

The **superior mesenteric artery** (1) gives off the following branches to the duodenum and colon:

2	**Inferior pancreaticoduodenal artery**, anastomoses via the superior, anterior and posterior pancreaticoduodenal arteries with the celiac trunk (→ 190)
3	**Jejunal arteries**
4	**Ileal arteries**
5	**Middle colic artery**
6	**Right colic artery**
7	**Ileocolic artery**
8	Colic branch, marginal arcade
9	**Anterior and posterior cecal arteries**
10	**Appendicular artery**

The arcades of the jejunal and ileal arteries run distant from the intestine. The rectal arteries, which communicate between the arcades and the intestine, are long. The arcades within the area of the colon lie close to the intestine, therefore the rectal arteries are short.

Location from which the superior mesenteric artery arises (→ 200)

8.6.3 Inferior Mesenteric Artery

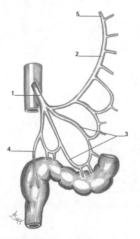

The inferior mesenteric artery (1) gives off the following branches:	
2	Left colic artery
3	Sigmoid arteries
4	Superior rectal artery (→ 192)

The structure labeled No. 5 is referred to as **Riolan's anastomosis**. It links the left colic artery with the middle colic artery and is located near the left colon flexure. At this point the areas supplied by the superior and inferior mesenteric artery anastomose.

In the same area also lies **Cannon's point**. Proximal to this position, the intestine is innervated by the parasympathetic vagus nerve and sympathetic superior mesenteric plexus. Distal to this point, it is innervated by the sacral parasympathetic nervous system and sympathetic inferior mesenteric plexus (→ 195).

Location from which the inferior mesenteric artery arises (→ 200)

8.6.4 Arterial Supply of Rectum

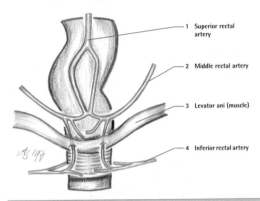

1 Superior rectal artery

2 Middle rectal artery

3 Levator ani (muscle)

4 Inferior rectal artery

The rectum is supplied by three arteries:		
1	Superior rectal artery	Arising from the inferior mesenteric artery (→ 191)
2	Middle rectal artery	Arising from the internal iliac artery (→ 215)
3	Inferior rectal artery	Arising from the internal pudendal artery, which in turn arises from the internal iliac artery (→ 215)

Again summarized below are the supplying vessels of each individual part of the intestine:		
Duodenum	Gastroduodenal artery	Arising from celiac trunk (→ 189)
Jejunum	Jejunal arteries	Arising from superior mesenteric artery (→ 190)
Ileum	Ileal arteries	
Cecum, ascending colon	Ileocolic artery, right colic artery	
Transverse colon	Middle colic artery	
Descending colon	Left colic artery	Arising from inferior mesenteric artery (→ 191)
Sigmoid colon	Sigmoid arteries	
Rectum	Superior, middle and inferior rectal arteries (→ 191),	see above

8.6.5 Portal Vein and its Tributaries

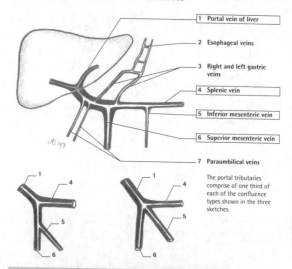

1 Portal vein of liver

2 Esophageal veins

3 Right and left gastric veins

4 Splenic vein

5 Inferior mesenteric vein

6 Superior mesenteric vein

7 Paraumbilical veins

The portal tributaries comprise of one third of each of the confluence types shown in the three sketches.

The veins of the abdominal organs run jointly with their corresponding arteries.

The **portal vein of liver**, which runs within the hepatoduodenal ligament, receives **tributaries from the unpaired abdominal organs**, the stomach, intestine (up to the upper third of the rectum), pancreas and spleen. The three main tributaries are the **splenic vein, superior mesenteric vein and the inferior mesenteric vein**. Blood from the liver drains via the **hepatic veins** into the inferior vena cava (→ 204).

The portal vein carries nutrient-rich, venous blood. The therein contained reabsorbed food components and substances are further processed or detoxified in the liver.

The portal venous system therefore consists of two consecutive capillary networks, those of the intestine and those of the liver.

8.6.6 Portal-systemic Anastomoses

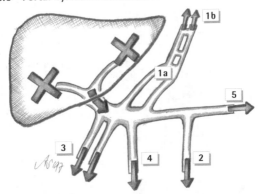

Clinical Information

Obstruction of blood flow in the portal vein of liver, e.g. due to a liver disease leading to cirrhosis, may cause a backup in supplying vessels. In this case the blood can still drain from the portal vein into the vena cava via **alternative collateral routes** called **portal-systemic anastomoses**.

The following alternate routes are available:

- **Backup of blood in the gastric veins (1a):** The blood drains by ways of anastomoses into the esophageal veins, then into the azygous vein and finally into the superior vena cava. When dilated, these are esophageal varices (1b, danger of bleeding!).
- **Backup of blood in the inferior mesenteric vein (2):** The blood flows via the anastomosing middle and inferior rectal vein into the iliac vein and finally into the inferior vena cava. When abnormally dilated these are **hemorrhoids.**
- **Backup of blood in the paraumbilical veins (3):** The blood drains via a recanalized umbilical vein (→ 157), and then via the veins of the skin (→ 124). These dilated veins are called caput medusae.

A backup of blood in the superior mesenteric vein (4) leads to non-specific symptoms (feeling of pressure etc.). A backup of blood in the splenic vein (5) causes splenic enlargement.

Liver cirrhosis is often accompanied by bleeding from esophageal varices. Loss of blood in this area is often severe and may be fatal.

8.6.7 Autonomic Nerve Supply of Abdominal Organs

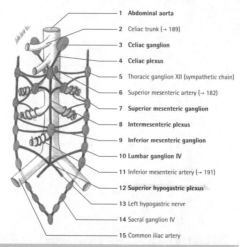

1 Abdominal aorta
2 Celiac trunk (→ 189)
3 Celiac ganglion
4 Celiac plexus
5 Thoracic ganglion XII (sympathetic chain)
6 Superior mesenteric artery (→ 182)
7 **Superior mesenteric ganglion**
8 Intermesenteric plexus
9 Inferior mesenteric ganglion
10 Lumbar ganglion IV
11 Inferior mesenteric artery (→ 191)
12 **Superior hypogastric plexus**
13 Left hypogastric nerve
14 Sacral ganglion IV
15 Common iliac artery

The **sympathetic nervous system** originates from the lateral horns of spinal cord segments Th1–L3. The greater splanchnic nerve arises from the 5th–9th thoracic ganglion, the lesser splanchnic nerve from the 9th–11th thoracic ganglion of the thoracic sympathetic trunk. Both nerves project to the celiac ganglion and to the celiac plexus. In addition, the thoracic sympathetic trunk sends off branches to supply the heart, lung and esophagus. Shown in the above figure is the abdominal and pelvic part of the sympathetic trunk. This area sends off branches to the abdominal organs.

Parasympathetic innervation of lungs, heart and gastrointestinal tract up to the Cannon point (→ 191) occurs via the **vagus nerve (X)**. The areas distal to the Cannon point receive innervation from nuclear areas of the sacral parasympathetic nervous system, that lie within spinal cord segments S2–S4.

Within the thoracic and abdominal space the sympathetic and parasympathetic nervous system form large plexuses. Examples are the **thoracic aortic plexus**, **abdominal aortic plexus** (= celiac plexus, No. 6 and 7 and smaller, not shown organ plexuses) and **superior hypogastric plexus**, from which smaller organ plexuses emanate.

8.6.8 Lymphatic System, Overview

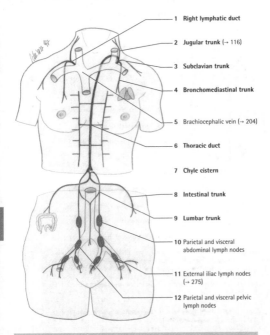

1 Right lymphatic duct

2 Jugular trunk (→ 116)

3 Subclavian trunk

4 Bronchomediastinal trunk

5 Brachiocephalic vein (→ 204)

6 Thoracic duct

7 Chyle cistern

8 Intestinal trunk

9 Lumbar trunk

10 Parietal and visceral
abdominal lymph nodes

11 External iliac lymph nodes
(→ 275)

12 Parietal and visceral pelvic
lymph nodes

Lymphatic drainage of the gastrointestinal tract is believed to be important for evacuating resorbed fats.

Refer to the indicated pages for more information on lymphatic drainage of the head-neck area (→ 116), the mammary (→ 230) and the inguinal region (→ 275).

9. Retroperitoneum

9.1 General Facts

9.1.1 Retroperitoneal Space, Overview

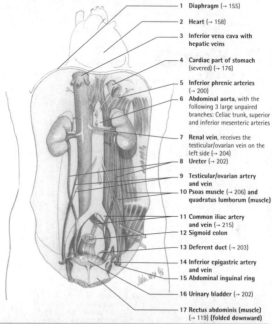

1 **Diaphragm** (→ 155)
2 **Heart** (→ 158)
3 **Inferior vena cava with hepatic veins**
4 **Cardiac part of stomach** (severed) (→ 176)
5 **Inferior phrenic arteries** (→ 200)
6 **Abdominal aorta**, with the following 3 large unpaired branches: Celiac trunk, superior and inferior mesenteric arteries
7 **Renal vein**, receives the testicular/ovarian vein on the left side (→ 204)
8 **Ureter** (→ 202)
9 **Testicular/ovarian artery and vein**
10 **Psoas muscle** (→ 206) and quadratus lumborum (muscle)
11 **Common iliac artery and vein** (→ 215)
12 **Sigmoid colon**
13 **Deferent duct** (→ 203)
14 **Inferior epigastric artery and vein**
15 **Abdominal inguinal ring**
16 **Urinary bladder** (→ 202)
17 **Rectus abdominis** (muscle) (→ 119) (folded downward)

Clinical Information

Aneurysms are outpouchings of the vessel wall. Eighty-five percent of aortic aneurysms are located infrarenal, 15 % occur in the thoracic area. To prevent potential rupture and resulting fatal bleeding, larger aneurysms must be repaired by surgery.

9.2 Urinary Apparatus, Suprarenal Glands

9.2.1 Kidneys

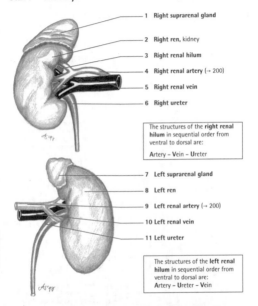

1 Right suprarenal gland

2 Right ren, kidney

3 Right renal hilum

4 Right renal artery (→ 200)

5 Right renal vein

6 Right ureter

> The structures of the **right renal hilum** in sequential order from ventral to dorsal are:
>
> Artery – **V**ein – **U**reter

7 Left suprarenal gland

8 Left ren

9 Left renal artery (→ 200)

10 Left renal vein

11 Left ureter

> The structures of the **left renal hilum** in sequential order from ventral to dorsal are:
>
> Artery – **U**reter – **V**ein

The **kidneys** are located **primarily retroperitoneal**. These organs are approximately 10 cm long and 5 cm wide. Their superior pole lies approximately at the level of the 12th thoracic vertebra, their inferior pole at the level of the 3rd lumbar vertebra. The left kidney normally lies **higher** than the right kidney. Together with the suprarenal glands, they are enclosed by the **adipose capsule**, which fixes them, although still moveable, in their position. The underlying tough **fibrous capsule** envelopes the kidney. The outer layer of the renal cortex is supplied by arteries of the adipose capsule and by terminal branches of the renal artery (→ 200).

9.2.2 Coronal Section of Kidney

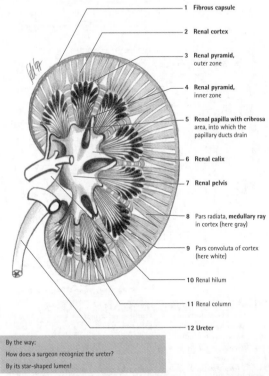

1 **Fibrous capsule**

2 **Renal cortex**

3 **Renal pyramid,** outer zone

4 **Renal pyramid,** inner zone

5 **Renal papilla with cribrosa** area, into which the papillary ducts drain

6 **Renal calix**

7 **Renal pelvis**

8 Pars radiata, **medullary ray** in cortex (here gray)

9 Pars convoluta of cortex (here white)

10 **Renal hilum**

11 **Renal column**

12 **Ureter**

By the way:

How does a surgeon recognize the ureter?

By its star-shaped lumen!

.2.3 Blood Supply of Kidney and Suprarenal Gland

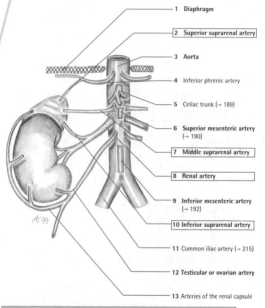

1 Diaphragm

2 Superior suprarenal artery

3 Aorta

4 Inferior phrenic artery

5 Celiac trunk (→ 189)

6 Superior mesenteric artery (→ 190)

7 Middle suprarenal artery

8 Renal artery

9 Inferior mesenteric artery (→ 192)

10 Inferior suprarenal artery

11 Common iliac artery (→ 215)

12 Testicular or ovarian artery

13 Arteries of the renal capsule

The suprarenal gland is supplied by three arteries:		
2	Superior suprarenal artery	Arising from the inferior phrenic artery
7	Middle suprarenal artery	Arising from the aorta
10	Inferior suprarenal artery	Arising from the renal artery
The kidney is supplied by three arteries:		
8	Renal artery	Arises from the aorta
13	Arteries of the renal capsule	Arise from the testicular/ovarian artery

9.2.4 Vascular Architecture of Kidney

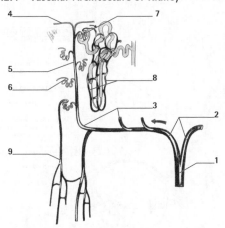

The **renal artery** branches at the renal hilum into **segmental arteries**. From these segmental arteries the **interlobar arteries** of the renal columns arise. The **branches and twigs of such a interlobar artery** (1) are shown in the above figure.

1	**Interlobar artery**
2	**Arcuate artery** (runs along the medulla-cortex border)
3	**Interlobular artery**
4	**Capsular branches of renal artery,** anastomoses with branches of capsular artery (→ 200)
5	**Afferent vessel** of glomerular loop
6	**Efferent vessel** of glomerular loop
7	Capillary loops in Bowman's capsule
8	Interstitial capillary network
9	**Arteriolae rectae**

9.2.5 Course and Narrowings of Ureter

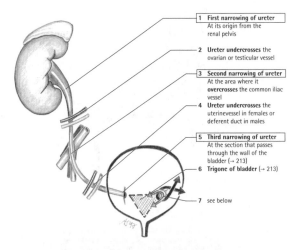

1 First narrowing of ureter
At its origin from the renal pelvis

2 Ureter undercrosses the ovarian or testicular vessel

3 Second narrowing of ureter
At the area where it **overcrosses** the common iliac vessel

4 Ureter undercrosses the uterinevessel in females or deferent duct in males

5 Third narrowing of ureter
At the section that passes through the wall of the bladder (→ 213)

6 Trigone of bladder (→ 213)

7 see below

To No. 5: Reflux of urine from the bladder back into the ureter is prevented by two mechanisms. First, the **intramural section of the ureter is positioned at an angle within the bladder wall**, which causes the ureter walls of this section to be pressed together when the bladder fills up. Second the ureteric orifice, located in the above shown trigone of bladder, is surrounded by muscular slings, which either open (dark arrows) or close (bright arrows) its lumen.

Clinical Information

In cases of lithiasis (stone formation within an organ), the narrowings of the ureter are of clinical relevance. The stones or stone fragments tend to lodge preferentially in these areas of decreased lumen width. The lodging is associated with severest colic pain. Available treatment options are ultrasonic (shock waves) fragmentation or endoscopic/operative removal.

9.3 Urogenital Tract of Male
9.3.1 Urogenital Tract of Male

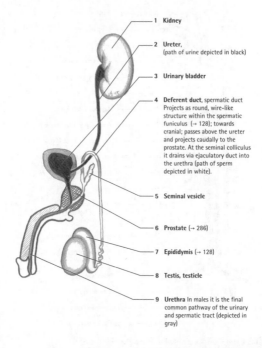

1 **Kidney**

2 **Ureter,**
(path of urine depicted in black)

3 **Urinary bladder**

4 **Deferent duct,** spermatic duct
Projects as round, wire-like
structure within the spermatic
funiculus (→ 128); towards
cranial; passes above the ureter
and projects caudally to the
prostate. At the seminal colliculus
it drains via ejaculatory duct into
the urethra (path of sperm
depicted in white).

5 **Seminal vesicle**

6 **Prostate** (→ 286)

7 **Epididymis** (→ 128)

8 **Testis, testicle**

9 **Urethra** In males it is the final
common pathway of the urinary
and spermatic tract (depicted in
gray)

9.4   Vessels, Nerves, Muscles

9.4.1   Veins of Dorsal Trunk Wall

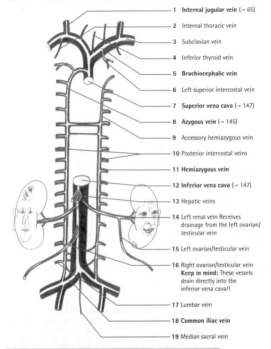

1   **Internal jugular vein** (→ 65)
2   Internal thoracic vein
3   Subclavian vein
4   Inferior thyroid vein
5   **Brachiocephalic vein**
6   Left superior intercostal vein
7   **Superior vena cava** (→ 147)
8   **Azygous vein** (→ 145)
9   Accessory hemiazygous vein
10   Posterior intercostal veins
11   **Hemiazygous vein**
12   **Inferior vena cava** (→ 147)
13   Hepatic veins
14   Left renal vein Receives drainage from the left ovarian/ testicular vein
15   Left ovarian/testicular vein
16   Right ovarian/testicular vein **Keep in mind: These vessels drain directly into the inferior vena cava!!**
17   Lumbar vein
18   **Common iliac vein**
19   Median sacral vein

See previous page regarding venous drainage of the abdominal organs (stomach, intestine, pancreas and spleen) (→ 193).
See previous page regarding the venous system of the anterior abdominal wall (→ 124)

9.4.2 Vascular Tracts of Retroperitoneal Space

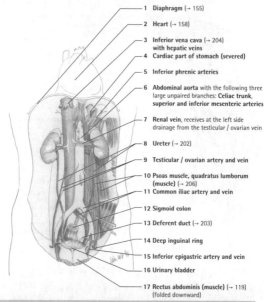

1 **Diaphragm** (→ 155)

2 **Heart** (→ 158)

3 **Inferior vena cava** (→ 204) with hepatic veins

4 **Cardiac part of stomach** (severed)

5 **Inferior phrenic arteries**

6 **Abdominal aorta** with the following three large unpaired branches: **Celiac trunk, superior and inferior mesenteric arteries**

7 **Renal vein**, receives at the left side drainage from the testicular / ovarian vein

8 **Ureter** (→ 202)

9 **Testicular / ovarian artery and vein**

10 **Psoas muscle, quadratus lumborum (muscle)** (→ 206)

11 **Common iliac artery and vein**

12 **Sigmoid colon**

13 **Deferent duct** (→ 203)

14 **Deep inguinal ring**

15 **Inferior epigastric artery and vein**

16 **Urinary bladder**

17 **Rectus abdominis (muscle)** (→ 119) (folded downward)

Course of the ureter: Undercrosses the ovarian vessel or testicular vessel, overcrosses the iliac vessel and finally undercrosses the deferent duct in males or the uterine vessel in females.

Remember: The three narrowings of the ureter are located at the renal pelvis origin, at the area overcrossing the common iliac vessel, and at the section that passes through the wall of the bladder.

Clinical Information

Nephrolithiasis is a condition in which stones moving down the ureter get stuck along the way. The narrowings of the ureter are especially prone to this kind of event.

9.4.3 Nerves and Muscles of Retroperitoneal Space

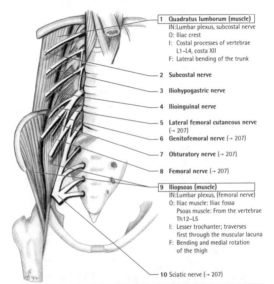

1 **Quadratus lumborum (muscle)**
IN: Lumbar plexus, subcostal nerve
O: Iliac crest
I: Costal processes of vertebrae
L1–L4, costa XII
F: Lateral bending of the trunk

2 **Subcostal nerve**

3 **Iliohypogastric nerve**

4 **Ilioinguinal nerve**

5 **Lateral femoral cutaneous nerve**
(→ 207)

6 **Genitofemoral nerve** (→ 207)

7 **Obturatory nerve** (→ 207)

8 **Femoral nerve** (→ 207)

9 **Iliopsoas (muscle)**
IN: Lumbar plexus, (femoral plexus)
O: Iliac muscle: Iliac fossa
Psoas muscle: From the vertebrae
Th12–L5
I: Lesser trochanter; traverses
first through the muscular lacuna
F: Bending and medial rotation
of the thigh

10 **Sciatic nerve** (→ 207)

The following nerves originate Lfrom the
lumbar plexus (→ 266)
- Iliohypogastric nerve
- Ilioinguinal nerve
- Genitofemoral nerve
- Lateral femoral cutaneous nerve
- Obturator nerve
- Femoral nerve

The bold initial letters of the words in this
mnemonic represent the initial letters of
the lumbar plexus nerves.

"Interested In Getting Laid On Fridays?"

The **femoral nerve** projects laterally to the
psoas muscle in a caudal direction through
the muscular lacuna.

The **obturator nerve** passes below the psoas
muscle and behind the internal iliac artery
into the obturator canal.

The **genitofemoral nerve** pierces the greater
psoas muscle.

The **sciatic nerve** arises from the sacral
plexus (→ 207)!

9.4.4 Lumbosacral Plexus

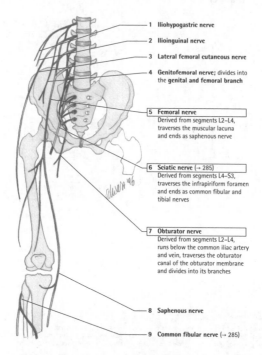

1 **Iliohypogastric nerve**

2 **Ilioinguinal nerve**

3 **Lateral femoral cutaneous nerve**

4 **Genitofemoral nerve;** divides into the **genital and femoral branch**

5 **Femoral nerve**
Derived from segments L2–L4, traverses the muscular lacuna and ends as saphenous nerve

6 **Sciatic nerve** (→ 285)
Derived from segments L4–S3, traverses the infrapiriform foramen and ends as common fibular and tibial nerves

7 **Obturator nerve**
Derived from segments L2–L4, runs below the common iliac artery and vein, traverses the obturator canal of the obturator membrane and divides into its branches

8 **Saphenous nerve**

9 **Common fibular nerve** (→ 285)

9.4.5 Lumbar Plexus, Innervation of Rectum and Bladder

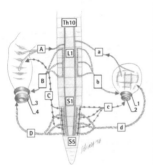

Innervation of the Rectum		
A	Th 11–S4	Rectal sensibility (pressure sensation)
B	L1–L3	Sympathetic supply of the internal anal sphincter muscle (3)
C	S2–S4	Parasympathetic supply of the rectal muscles
D	S2–S4	Motor supply of external anal sphincter muscle (4); voluntary!
Innervation of the Bladder		
a	Th11–S4	Sensory supply of bladder (pressure sensation)
b	L1–L3	Sympathetic supply of internal sphincter (1)
c	S2–S4	Parasympathetic supply of bladder muscles
d	S2–S4	Motor supply of external sphincter (2); voluntary!

Clinical Information

Under normal conditions, urination and defecation are initiated when the inner pressure reaches a threshold (**"reflex rectum"**, **"reflex bladder"**). In **paraplegia** of the area above Th11, the patient is not able to empty the intestine and bladder voluntarily, partly due to the loss of pressure sensibility.

Paraplegia of the area below the sympathetic and above the parasympathetic segments from which they originate, leads to an **"overflow rectum"** and/or an **"overflow bladder"** – emptying occurs when the inner pressure overcomes the closing pressure of the internal sphincter.

10. Pelvis

10.1 Foramina, Muscles, Vessels, Ligaments

10.1.1 Greater and Lesser Sciatic Foramen

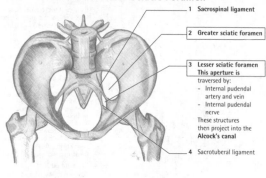

1 Sacrospinal ligament

2 Greater sciatic foramen

3 Lesser sciatic foramen
This aperture is
traversed by:
- Internal pudendal artery and vein
- Internal pudendal nerve

These structures then project into the **Alcock's canal**

4 Sacrotuberal ligament

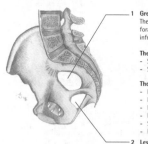

1 **Greater sciatic foramen**
The piriform muscle (→ 214) divides this foramen into a supra- and infrapiriform part.

The suprapiriform foramen is traversed by:
- Superior gluteal artery and vein (→ 285)
- Superior gluteal nerve (→ 257)

The infrapiriform foramen is traversed by:
- Inferior gluteal artery and vein (→ 285)
- Inferior gluteal nerve (→ 257)
- Internal pudendal artery and vein
- Internal pudendal nerve (→ 285)
- Sciatic nerve (→ 285)
- Posterior femoral cutaneous nerve (→ 301)

2 Lesser sciatic foramen

10.1.2 Musculature of Pelvic Floor, Cranial View

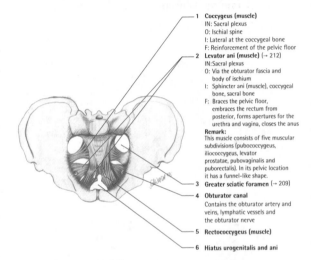

1 **Coccygeus (muscle)**
   IN: Sacral plexus
   O: Ischial spine
   I: Lateral at the coccygeal bone
   F: Reinforcement of the pelvic floor

2 **Levator ani (muscle)** (→ 212)
   IN: Sacral plexus
   O: Via the obturator fascia and body of ischium
   I: Sphincter ani (muscle), coccygeal bone, sacral bone
   F: Braces the pelvic floor, embraces the rectum from posterior, forms apertures for the urethra and vagina, closes the anus
   Remark:
   This muscle consists of five muscular subdivisions (pubococcygeus, iliococcygeus, levator prostatae, pubovaginalis and puborectalis). In its pelvic location it has a funnel-like shape.

3 **Greater sciatic foramen** (→ 209)

4 **Obturator canal**
   Contains the obturator artery and veins, lymphatic vessels and the obturator nerve

5 **Rectococcygeus (muscle)**

6 **Hiatus urogenitalis and ani**

The pelvic floor can be divided into several compartments:	
Compartment	**Muscles**
Pelvic diaphragm	- Levator ani (muscle)
	- Coccygeus (muscle)
Urogenital diaphragm	- Deep and superficial transverse muscles of perineum
	- Deep transverse muscle of perineum
In addition	- Sphincter urethrae and sphincter ani externus (muscle)
	- Ischiocavernous muscle
	- Bulbospongiosus (muscle)

10.1.3 Musculature of Pelvic Floor, View from Below

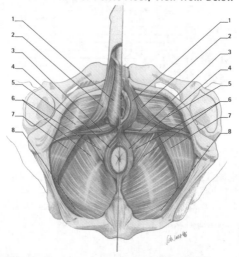

Floor of the male pelvis

Floor of the female pelvis

1	Bulbospongiosus (muscle) (left: Surrounds the spongy body of penis – right: Passes around the vaginal and urethral orifice)
2	Ischiocavernous muscles
3	Superficial transverse muscle of perineum
4	Sphincter ani externus (muscle)
5	Deep transverse muscle of perineum
6	Levator ani (muscle) (→ 212)
7	Obturator fascia
8	Coccygeus (muscle)

10.1.4 Levator Ani, Small Pelvis of Female

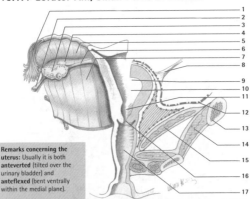

Remarks concerning the uterus: Usually it is both **anteverted** (tilted over the urinary bladder) and **anteflexed** (bent ventrally within the medial plane).

1 Uterine tubes with fimbriae (open connection to the abdominal cavity!), ampulla and isthmus
2 Suspensory ligament of ovary (with ovarian arteries and veins)
3 Ovary
4 Mesovarium
5 Mesosalpinx
6 Proper ligament of ovary, connects the ovary with the lateral angle of uterus
7 Broad ligament of uterus, contains vessels, nerves travelling to the uterus, ureter
8 Round ligament of uterus, holds the uterus in place. Projects from the lateral angle of uterus through the broad ligament, then passes through the inguinal canal and ends in the labia majora (→ 127)
   Parts of the **Uterus:**
9 Fundus of uterus
10 Corpus of uterus
11 Cervix of uterus
12 Peritoneum
13 Superior ramus of pubis
14 Internal obturator muscle (→ 214)
15 Levator ani (muscle)
16 Opening of uterus
17 Vagina

To 15 Levator ani (muscle) (→ 211)
IN: Branches from the sacral plexus (SIII–SIV)
O: Tendinous arch of levator ani (muscle)
I: Sacral bone, coccygeal bone (radiates into the sphincter ani externus)
F: This muscle forms a muscular sling, which embraces the rectum from behind. The medial free edge leaves a gap for the urethra and in females also for the vagina (urogenital hiatus)

10.1.5 Frontal Sections Through Pelvis

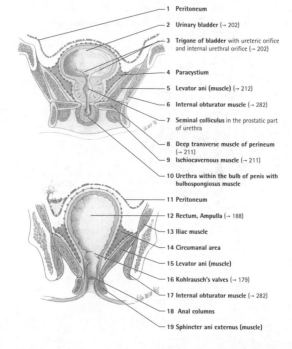

1 Peritoneum

2 Urinary bladder (→ 202)

3 Trigone of bladder with ureteric orifice and internal urethral orifice (→ 202)

4 Paracystium

5 Levator ani (muscle) (→ 212)

6 Internal obturator muscle (→ 282)

7 Seminal colliculus in the prostatic part of urethra

8 Deep transverse muscle of perineum (→ 211)

9 Ischiocavernous muscle (→ 211)

10 Urethra within the bulb of penis with bulbospongiosus muscle

11 Peritoneum

12 Rectum, Ampulla (→ 188)

13 Iliac muscle

14 Circumanal area

15 Levator ani (muscle)

16 Kohlrausch's valves (→ 179)

17 Internal obturator muscle (→ 282)

18 Anal columns

19 Sphincter ani externus (muscle)

10.1.6 Piriformis and Internal Obturator Muscles

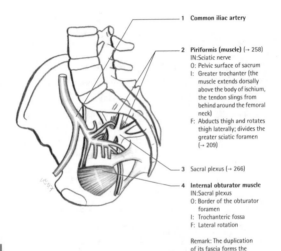

1 **Common iliac artery**

2 **Piriformis (muscle)** (→ 258)
IN: Sciatic nerve
O: Pelvic surface of sacrum
I: Greater trochanter (the
muscle extends dorsally
above the body of ischium,
the tendon slings from
behind around the femoral
neck)
F: Abducts thigh and rotates
thigh laterally; divides the
greater sciatic foramen
(→ 209)

3 Sacral plexus (→ 266)

4 **Internal obturator muscle**
IN: Sacral plexus
O: Border of the obturator
foramen
I: Trochanteric fossa
F: Lateral rotation

Remark: The duplication
of its fascia forms the
Alcock's canal
(→ 286)

10.1.7 Internal Iliac Artery

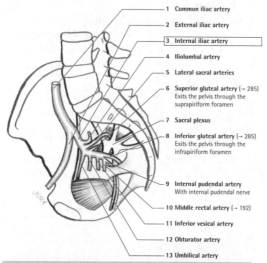

1 Common iliac artery

2 External iliac artery

3 Internal iliac artery

4 Iliolumbal artery

5 Lateral sacral arteries

6 Superior gluteal artery (→ 285)
Exits the pelvis through the
suprapiriform foramen

7 Sacral plexus

8 Inferior gluteal artery (→ 285)
Exits the pelvis through the
infrapiriform foramen

9 Internal pudendal artery
With internal pudendal nerve

10 Middle rectal artery (→ 192)

11 Inferior vesical artery

12 Obturator artery

13 Umbilical artery

The **umbilical artery** gives rise to the **superior vesical artery** and to the **artery of deferent duct or uterine artery**.

The internal pudendal artery gives rise to branches that project to the penis or to the clitoris and vestibule of vagina. It also gives rise to the **inferior rectal artery**. The internal pudendal artery exits the pelvis through the infrapiriform part of the greater sciatic foramen and passes together with the internal pudendal nerve back into the pelvis through the lesser sciatic foramen. From here it projects within the Alcock's canal (→ 286) towards the symphysis.

The very variable **branches of the internal iliac artery** are divided into **visceral** branches (superior and inferior vesical arteries, uterine artery, artery of deferent duct, middle and inferior rectal arteries) and **parietal** branches (all others).

The **external iliac artery** passes below the inguinal ligament through the vascular lacuna (→ 262) and transitions into the femoral artery

11.   Upper Limb

11.1   Shoulder

11.1.1   Shoulder: Bones and Ligaments

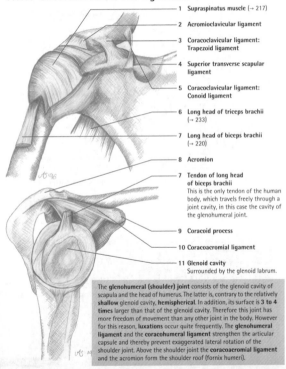

1 Supraspinatus muscle (→ 217)

2 Acromioclavicular ligament

3 Coracoclavicular ligament: Trapezoid ligament

4 Superior transverse scapular ligament

5 Coracoclavicular ligament: Conoid ligament

6 Long head of triceps brachii (→ 233)

7 Long head of biceps brachii (→ 220)

8 Acromion

7 Tendon of long head of biceps brachii
This is the only tendon of the human body, which travels freely through a joint cavity, in this case the cavity of the glenohumeral joint.

9 Coracoid process

10 Coracoacromial ligament

11 Glenoid cavity
Surrounded by the glenoid labrum.

The **glenohumeral (shoulder) joint** consists of the glenoid cavity of scapula and the head of humerus. The latter is, contrary to the relatively **shallow** glenoid cavity, **hemispherical**. In addition, its surface is **3 to 4 times** larger than that of the glenoid cavity. Therefore this joint has more freedom of movement than any other joint in the body. However for this reason, **luxations** occur quite frequently. The **glenohumeral ligament** and the **coracohumeral ligament** strengthen the articular capsule and thereby prevent exaggerated lateral rotation of the shoulder joint. Above the shoulder joint the **coracoacromial ligament** and the acromion form the shoulder roof (fornix humeri).

11.1.2 Musculotendinous Rotator Cuff

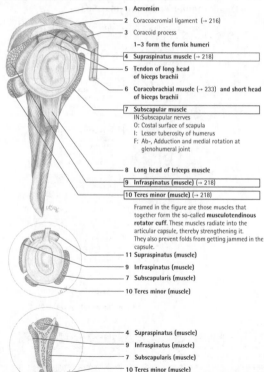

1 Acromion

2 Coracoacromial ligament (→ 216)

3 Coracoid process

1–3 form the fornix humeri

4 **Supraspinatus muscle** (→ 218)

5 Tendon of long head of biceps brachii

6 **Coracobrachial muscle** (→ 233) **and short head of biceps brachii**

7 **Subscapular muscle**
IN: Subscapular nerves
O: Costal surface of scapula
I: Lesser tuberosity of humerus
F: Ab-, Adduction and medial rotation at glenohumeral joint

8 **Long head of triceps muscle**

9 **Infraspinatus (muscle)** (→ 218)

10 **Teres minor (muscle)** (→ 218)

Framed in the figure are those muscles that together form the so-called **musculotendinous rotator cuff**. These muscles radiate into the articular capsule, thereby strengthening it. They also prevent folds from getting jammed in the capsule.

11 Supraspinatus (muscle)

9 Infraspinatus (muscle)

7 Subscapularis (muscle)

10 Teres minor (muscle)

4 Supraspinatus (muscle)

9 Infraspinatus (muscle)

7 Subscapularis (muscle)

10 Teres minor (muscle)

11.1.3 Musculotendinous Rotator Cuff (In Situ), Quadrangular and Triangular Space

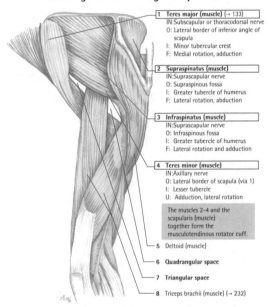

1 Teres major (muscle) (→ 133)
IN: Subscapular or thoracodorsal nerve
O: Lateral border of inferior angle of scapula
I: Minor tubercular crest
F: Medial rotation, adduction

2 Supraspinatus (muscle)
IN: Suprascapular nerve
O: Supraspinous fossa
I: Greater tubercle of humerus
F: Lateral rotation, abduction

3 Infraspinatus (muscle)
IN: Suprascapular nerve
O: Infraspinous fossa
I: Greater tubercle of humerus
F: Lateral rotation and adduction

4 Teres minor (muscle)
IN: Axillary nerve
O: Lateral border of scapula (via 1)
I: Lesser tubercle
U: Adduction, lateral rotation

The muscles 2–4 and the scapularis (muscle) together form the musculotendinous rotator cuff.

5 Deltoid (muscle)

6 Quadrangular space

7 Triangular space

8 Triceps brachii (muscle) (→ 232)

Space	Boundaries	Traversing structures
Triangular space	Teres major and minor (muscle), triceps brachii (muscle),	Scapular circumflex artery and vein (→ 220)
Quadrangula space	Teres major and minor (muscle), triceps brachii (muscle), humerus	Axillary nerve, posterior humeral circumflex artery and vein (→ 225)

11.1.4 Arm, Shoulder: Muscles and Radial Nerve

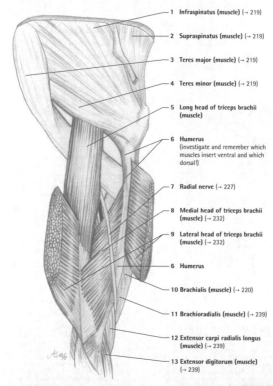

1 **Infraspinatus (muscle)** (→ 219)

2 **Supraspinatus (muscle)** (→ 219)

3 **Teres major (muscle)** (→ 219)

4 **Teres minor (muscle)** (→ 219)

5 **Long head of triceps brachii (muscle)**

6 **Humerus** (investigate and remember which muscles insert ventral and which dorsal!)

7 **Radial nerve** (→ 227)

8 **Medial head of triceps brachii (muscle)** (→ 232)

9 **Lateral head of triceps brachii (muscle)** (→ 232)

6 **Humerus**

10 **Brachialis (muscle)** (→ 220)

11 **Brachioradialis (muscle)** (→ 239)

12 **Extensor carpi radialis longus (muscle)** (→ 239)

13 **Extensor digitorum (muscle)** (→ 239)

11.1.5 Arm, Shoulder: Muscles, Vessels, Nerves (Ventral View)

1 Levator scapulae (muscle)

2 Subscapularis (muscle) (→ 231)

3 Rhomboid muscle (→ 132)

4 Subclavius (muscle and coracoclavicular ligament

5 Coracobrachialis (muscle)

6 Subscapular artery (→ 224)

7 Scapular circumflex artery (→ 224)

8 Thoracodorsal artery (→ 224)

9 Ulnar nerve (see below)

10 Median nerve (see below)

11 Brachial artery (→ 225)
More distally it branches off the superior and inferior ulnar collateral arteries!

12 Biceps brachii (muscle)
IN: Musculocutaneous nerve
O: Long head (→ 216) Supraglenoid tubercle of scapula,
Short head: Coracoid process of scapula
I: Tuberosity of radius, bicipital aponeurosis
F: Anteversion and medial rotation of shoulders, abduction (long head) and adduction (short head). Flexion and supination at elbow

13 Brachialis (muscle)
IN: Musculocutaneous nerve
O: Ventral aspect of humerus
I: Tuberosity of ulna
F: Flexes forearm in all positionsr

14 Triceps brachii (muscles) (→ 232)

The ulnar nerve (→ 242) projects out of the axilla, medial to the axillary artery, in a dorsal direction through the medial brachial intermuscular septum. It crosses, posterior to the medial humeral epicondyle, the elbow-joint, continues in a distal direction below the flexor carpi ulnaris (muscle) on the volar side, and divides at the flexor retinaculum into its terminal branches. The median nerve projects from the medial bicipital fissure (→ 242) above the brachial artery towards medial, passes the elbow-joint below the biceps brachii (muscle) and above the brachialis (muscle), pierces the pronator teres (muscle), projects between the superficial and profound flexor digitorum (muscle) in a distal direction, and travels below the flexor retinaculum to the hand. Carpal tunnel syndrome results from compression of the median nerve within the carpal tunnel.

11.1.6 Muscle Attachments of Scapula

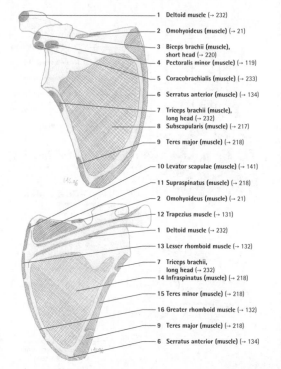

1 Deltoid muscle (→ 232)
2 Omohyoideus (muscle) (→ 21)
3 Biceps brachii (muscle), short head (→ 220)
4 Pectoralis minor (muscle) (→ 119)
5 Coracobrachialis (muscle) (→ 233)
6 Serratus anterior (muscle) (→ 134)
7 Triceps brachii (muscle), long head (→ 232)
8 Subscapularis (muscle) (→ 217)
9 Teres major (muscle) (→ 218)

10 Levator scapulae (muscle) (→ 141)
11 Supraspinatus (muscle) (→ 218)
2 Omohyoideus (muscle) (→ 21)
12 Trapezius muscle (→ 131)
1 Deltoid muscle (→ 232)
13 Lesser rhomboid muscle (→ 132)
7 Triceps brachii, long head (→ 232)
14 Infraspinatus (muscle) (→ 218)
15 Teres minor (muscle) (→ 218)
16 Greater rhomboid muscle (→ 132)
9 Teres major (muscle) (→ 218)
6 Serratus anterior (muscle) (→ 134)

11.2 Axillary Region

11.2.1 Boundaries and Fascia of Axilla

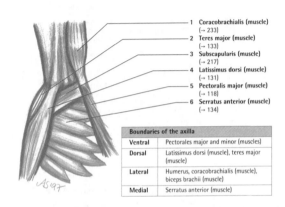

1 Coracobrachialis (muscle)
  (→ 233)
2 Teres major (muscle)
  (→ 133)
3 Subscapularis (muscle)
  (→ 217)
4 Latissimus dorsi (muscle)
  (→ 131)
5 Pectoralis major (muscle)
  (→ 118)
6 Serratus anterior (muscle)
  (→ 134)

Boundaries of the axilla	
Ventral	Pectorales major and minor (muscles)
Dorsal	Latissimus dorsi (muscle), teres major (muscle)
Lateral	Humerus, coracobrachialis (muscle), biceps brachii (muscle)
Medial	Serratus anterior (muscle)

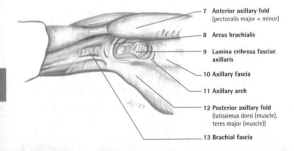

7 Anterior axillary fold
  (pectoralis major + minor)

8 Arcus brachialis

9 Lamina cribrosa fasciae
  axillaris

10 Axillary fascia

11 Axillary arch

12 Posterior axillary fold
   (latissimus dorsi (muscle),
   teres major (muscle))

13 Brachial fascia

11.2.2 Quadrangular and Triangular Space (Dorsal View)

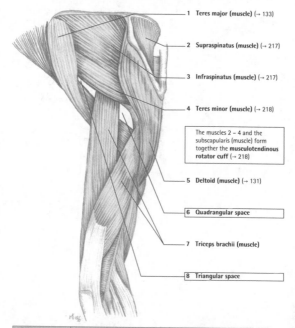

1 Teres major (muscle) (→ 133)

2 Supraspinatus (muscle) (→ 217)

3 Infraspinatus (muscle) (→ 217)

4 Teres minor (muscle) (→ 218)

The muscles 2 – 4 and the subscapularis (muscle) form together the **musculotendinous rotator cuff** (→ 218)

5 Deltoid (muscle) (→ 131)

6 Quadrangular space

7 Triceps brachii (muscle)

8 Triangular space

Space	Boundaries	Traversing structures
Triangular space	Teres major and minor (muscle), triceps brachii (muscle),	Scapular circumflex artery and vein
Quadrangular space	Teres major and minor (muscle), triceps brachii (muscle), humerus	Axillary nerve, posterior humeral circumflex artery and vein

11.2.3 Arteries of Axillary Region

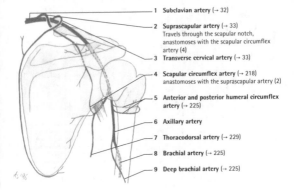

1 **Subclavian artery** (→ 32)

2 **Suprascapular artery** (→ 33)
Travels through the scapular notch,
anastomoses with the scapular circumflex
artery (4)

3 **Transverse cervical artery** (→ 33)

4 **Scapular circumflex artery** (→ 218)
anastomoses with the suprascapular artery (2)

5 **Anterior and posterior humeral circumflex
artery** (→ 225)

6 **Axillary artery**

7 **Thoracodorsal artery** (→ 229)

8 **Brachial artery** (→ 225)

9 **Deep brachial artery** (→ 225)

11.2.4 Arteries of Upper Limb

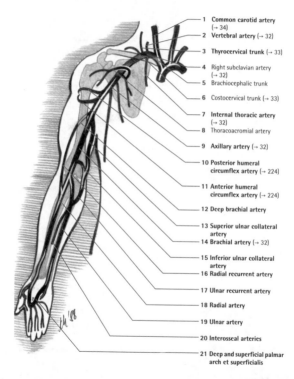

1 **Common carotid artery** (→ 34)
2 **Vertebral artery** (→ 32)
3 **Thyrocervical trunk** (→ 33)
4 Right subclavian artery (→ 32)
5 Brachiocephalic trunk
6 Costocervical trunk (→ 33)
7 **Internal thoracic artery** (→ 32)
8 Thoracoacromial artery
9 **Axillary artery** (→ 32)
10 Posterior humeral circumflex artery (→ 224)
11 Anterior humeral circumflex artery (→ 224)
12 Deep brachial artery
13 Superior ulnar collateral artery
14 Brachial artery (→ 32)
15 Inferior ulnar collateral artery
16 Radial recurrent artery
17 Ulnar recurrent artery
18 Radial artery
19 Ulnar artery
20 Interosseal arteries
21 Deep and superficial palmar arch et superficialis

11.2.5 Nerves of Arm and Lateral Wall of Thorax

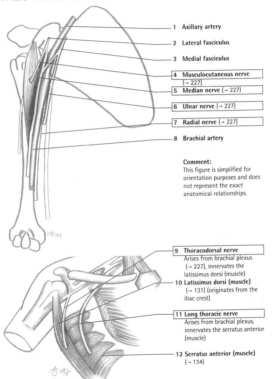

1 Axillary artery

2 Lateral fasciculus

3 Medial fasciculus

4 **Musculocutaneous nerve** (→ 227)

5 **Median nerve** (→ 227)

6 **Ulnar nerve** (→ 227)

7 **Radial nerve** (→ 227)

8 Brachial artery

Comment:
This figure is simplified for orientation purposes and does not represent the exact anatomical relationships.

9 **Thoracodorsal nerve**
Arises from brachial plexus (→ 227), innervates the latissimus dorsi (muscle)

10 **Latissimus dorsi (muscle)** (→ 131) (originates from the iliac crest)

11 **Long thoracic nerve**
Arises from brachial plexus, innervates the serratus anterior (muscle)

12 **Serratus anterior (muscle)** (→ 134)

11.2.6 Brachial Plexus In Situ

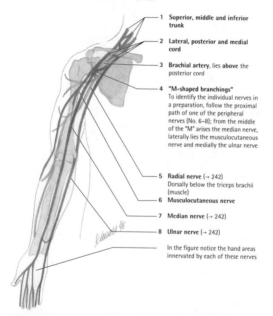

1 Superior, middle and inferior trunk

2 Lateral, posterior and medial cord

3 Brachial artery, lies **above** the posterior cord

4 "M-shaped branchings"
To identify the individual nerves in a preparation, follow the proximal path of one of the peripheral nerves (No. 6–8); from the middle of the "M" arises the median nerve, laterally lies the musculocutaneous nerve and medially the ulnar nerve

5 Radial nerve (→ 242)
Dorsally below the triceps brachii (muscle)

6 Musculocutaneous nerve

7 Median nerve (→ 242)

8 Ulnar nerve (→ 242)

In the figure notice the hand areas innervated by each of these nerves

Upper limb muscle		Innervation
Deltoid muscle	(→ 232)	Axillary nerve
Biceps brachii (muscle)	(→ 220)	Musculocutaneous nerve
Coracobrachialis (muscle)	(→ 233)	Musculocutaneous nerve
Brachialis (muscle)	(→ 220)	Musculocutaneous nerve
Triceps brachii (muscle)	(→ 232)	Radial nerve

11.2.7 Brachial Plexus (Schematic)

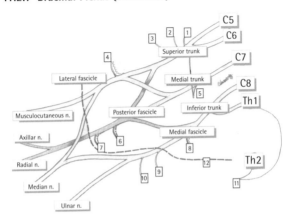

Below are the nerves labeled 1 through 11	
1	Dorsal scapular nerve
2	Suprascapular nerve
3	Subclavius nerve
4	Lateral pectoral nerve
5	Long thoracic nerve
6	Subscapular nerve
7	Thoracodorsal nerve
8	Medial pectoral nerve
9	Medial cutaneous nerve of arm
10	Lateral cutaneous nerve of arm
11	Intercostobrachial nerve
(12)	(borderline between the supra- and infraclavicular part)

11.2.8 Vessels of Lateral Trunk Wall, Transverse Section of Arm

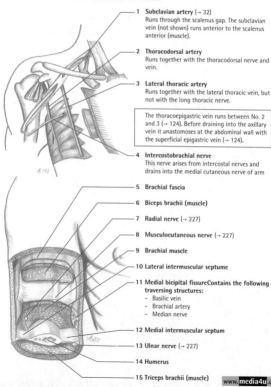

1 **Subclavian artery** (→ 32)
Runs through the scalenus gap. The subclavian vein (not shown) runs anterior to the scalenus anterior (muscle).

2 **Thoracodorsal artery**
Runs together with the thoracodorsal nerve and vein.

3 **Lateral thoracic artery**
Runs together with the lateral thoracic vein, but not with the long thoracic nerve.

> The thoracoepigastric vein runs between No. 2 and 3 (→ 124). Before draining into the axillary vein it anastomoses at the abdominal wall with the superficial epigastric vein (→ 124).

4 **Intercostobrachial nerve**
This nerve arises from intercostal nerves and drains into the medial cutaneous nerve of arm

5 **Brachial fascia**

6 **Biceps brachii (muscle)**

7 **Radial nerve** (→ 227)

8 **Musculocutaneous nerve** (→ 227)

9 **Brachial muscle**

10 **Lateral intermuscular septume**

11 **Medial bicipital fissure** Contains the following traversing structures:
   - Basilic vein
   - Brachial artery
   - Median nerve

12 **Medial intermuscular septum**

13 **Ulnar nerve** (→ 227)

14 **Humerus**

15 **Triceps brachii (muscle)**

11.2.9 Lymphatic Drainage Ways of Female Breast

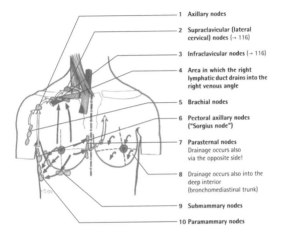

1 Axillary nodes

2 Supraclavicular (lateral cervical) nodes (→ 116)

3 Infraclavicular nodes (→ 116)

4 Area in which the right lymphatic duct drains into the right venous angle

5 Brachial nodes

6 Pectoral axillary nodes ("Sorgius node")

7 Parasternal nodes
Drainage occurs also via the opposite side!

8 Drainage occurs also into the deep interior (bronchomediastinal trunk)

9 Submammary nodes

10 Paramammary nodes

Direction of drainage is indicated in the figure by arrows. The final path of the left side drains together with the thoracic duct (→ 196) into the left venous angle. Approximately 75% of lymphatic drainage from the female breast into the axilla occurs via No. 6 and 1. A small amount drains via No. 7.

Clinical Information

The lymphatic drainage ways of the female breast are clinically relevant due to their involvement in lymphogenic metastasis of breast cancer. In females this is the most frequently observed type of cancer. The left breast is more prone to being affected and in approximately 60% of cases the cancer is located in the upper outer quadrant. Due to the possibility of lymphogenic metastasis, removal of the axillary lymph nodes may be indicated depending on the stage of the disease.

11.3 Upper Arm

11.3.1 Muscle origins and Insertions of Humerus

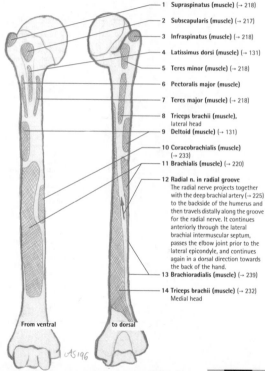

1 **Supraspinatus (muscle)** (→ 218)

2 **Subscapularis (muscle)** (→ 217)

3 **Infraspinatus (muscle)** (→ 218)

4 **Latissimus dorsi (muscle)** (→ 131)

5 **Teres minor (muscle)** (→ 218)

6 **Pectoralis major (muscle)**

7 **Teres major (muscle)** (→ 218)

8 **Triceps brachii (muscle),** lateral head

9 **Deltoid (muscle)** (→ 131)

10 **Coracobrachialis (muscle)** (→ 233)

11 **Brachialis (muscle)** (→ 220)

12 **Radial n. in radial groove**
The radial nerve projects together with the deep brachial artery (→ 225) to the backside of the humerus and then travels distally along the groove for the radial nerve. It continues anteriorly through the lateral brachial intermuscular septum, passes the elbow joint prior to the lateral epicondyle, and continues again in a dorsal direction towards the back of the hand.

13 **Brachioradialis (muscle)** (→ 239)

14 **Triceps brachii (muscle)** (→ 232) Medial head

From ventral

to dorsal

11.3.2 Muscles of Upper Arm (Dorsal and Lateral View)

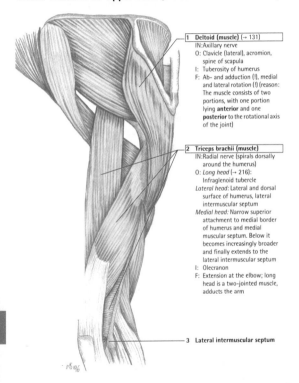

1 Deltoid (muscle) (→ 131)
IN: Axillary nerve
O: Clavicle (lateral), acromion, spine of scapula
I: Tuberosity of humerus
F: Ab- and adduction (!), medial and lateral rotation (!) (reason: The muscle consists of two portions, with one portion lying **anterior** and one **posterior** to the rotational axis of the joint)

2 Triceps brachii (muscle)
IN: Radial nerve (spirals dorsally around the humerus)
O: *Long head* (→ 216): Infraglenoid tubercle
Lateral head: Lateral and dorsal surface of humerus, lateral intermuscular septum
Medial head: Narrow superior attachment to medial border of humerus and medial muscular septum. Below it becomes increasingly broader and finally extends to the lateral intermuscular septum
I: Olecranon
F: Extension at the elbow; long head is a two-jointed muscle, adducts the arm

3 Lateral intermuscular septum

11.3.3 Biceps Brachii, Coracobrachialis

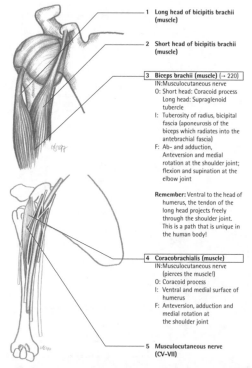

1 **Long head of bicipitis brachii (muscle)**

2 **Short head of bicipitis brachii (muscle)**

3 **Biceps brachii (muscle) (→ 220)**
IN: Musculocutaneous nerve
O: Short head: Coracoid process
Long head: Supraglenoid tubercle
I: Tuberosity of radius, bicipital fascia (aponeurosis of the biceps which radiates into the antebrachial fascia)
F: Ab- and adduction, Anteversion and medial rotation at the shoulder joint; flexion and supination at the elbow joint

Remember: Ventral to the head of humerus, the tendon of the long head projects freely through the shoulder joint. This is a path that is unique in the human body!

4 **Coracobrachialis (muscle)**
IN: Musculocutaneous nerve (pierces the muscle!)
O: Coracoid process
I: Ventral and medial surface of humerus
F: Anteversion, adduction and medial rotation at the shoulder joint

5 **Musculocutaneous nerve (CV-VII)**

11.3.4 Muscle Septa of Upper Arm

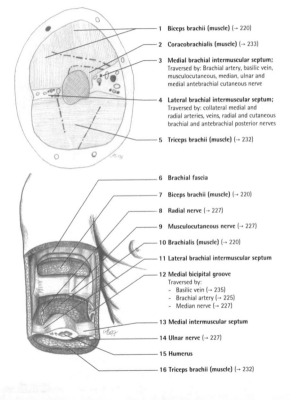

1 **Biceps brachii (muscle)** (→ 220)

2 **Coracobrachialis (muscle)** (→ 233)

3 **Medial brachial intermuscular septum;**
Traversed by: Brachial artery, basilic vein, musculocutaneous, median, ulnar and medial antebrachial cutaneous nerve

4 **Lateral brachial intermuscular septum;**
Traversed by: collateral medial and radial arteries, veins, radial and cutaneous brachial and antebrachial posterior nerves

5 **Triceps brachii (muscle)** (→ 232)

6 **Brachial fascia**

7 **Biceps brachii (muscle)** (→ 220)

8 **Radial nerve** (→ 227)

9 **Musculocutaneous nerve** (→ 227)

10 **Brachialis (muscle)** (→ 220)

11 **Lateral brachial intermuscular septum**

12 **Medial bicipital groove**
Traversed by:
- Basilic vein (→ 235)
- Brachial artery (→ 225)
- Median nerve (→ 227)

13 **Medial intermuscular septum**

14 **Ulnar nerve** (→ 227)

15 **Humerus**

16 **Triceps brachii (muscle)** (→ 232)

11.3.5 Superficial Veins of Arm

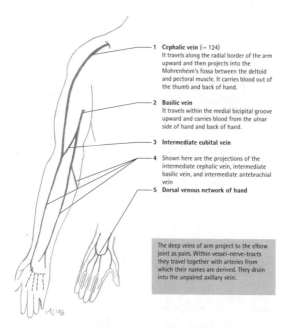

1 **Cephalic vein** (→ 124)
It travels along the radial border of the arm upward and then projects into the Mohrenheim's fossa between the deltoid and pectoral muscle. It carries blood out of the thumb and back of hand.

2 **Basilic vein**
It travels within the medial bicipital groove upward and carries blood from the ulnar side of hand and back of hand.

3 **Intermediate cubital vein**

4 Shown here are the projections of the intermediate cephalic vein, intermediate basilic vein, and intermediate antebrachial vein

5 **Dorsal venous network of hand**

The deep veins of arm project to the elbow joint as pairs. Within vessel-nerve-tracts they travel together with arteries from which their names are derived. They drain into the unpaired axillary vein.

11.4 Forearm

11.4.1 Ligaments of Cubital Joint

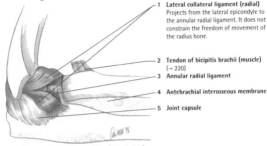

1 **Lateral collateral ligament (radial)**
Projects from the lateral epicondyle to the annular radial ligament. It does not constrain the freedom of movement of the radius bone.

2 **Tendon of bicipitis brachii (muscle)** (→ 220)

3 **Annular radial ligament**

4 **Antebrachial interosseous membrane**

5 **Joint capsule**

Forearm in pronation position (lateral view)

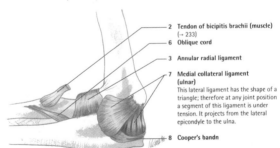

2 **Tendon of bicipitis brachii (muscle)** (→ 233)

6 **Oblique cord**

3 **Annular radial ligament**

7 **Medial collateral ligament (ulnar)**
This lateral ligament has the shape of a triangle; therefore at any joint position a segment of this ligament is under tension. It projects from the lateral epicondyle to the ulna.

8 **Cooper's bandn**

Forearm in pronation position (medial view)

The annular radial ligament embraces the radial head; it arises from and inserts into the ulna; this enables the radial head to freely rotate about its longitudinal axis. The annular ligament is completely integrated into the joint capsule. When making an anatomical preparation it is therefore necessary to create artificial boundaries in order to display it in a defined way (the joint capsule must be completely removed anyway). The movement possibilities of the bones as well as of the elbow joint – flexion, extension, pronation and supination – can then easily be reproduced.

11.4.2 Muscle origins and Insertions of Forearm (Dorsal View)

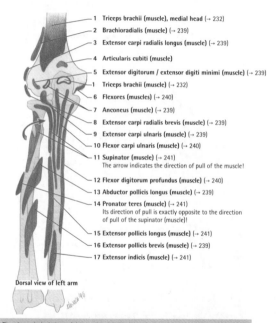

1 Triceps brachii (muscle), medial head (→ 232)

2 Brachioradialis (muscle) (→ 239)

3 Extensor carpi radialis longus (muscle) (→ 239)

4 Articularis cubiti (muscle)

5 Extensor digitorum / extensor digiti minimi (muscle) (→ 239)

1 Triceps brachii (muscle) (→ 232)

6 Flexores (muscles) (→ 240)

7 Anconeus (muscle) (→ 239)

8 Extensor carpi radialis brevis (muscle) (→ 239)

9 Extensor carpi ulnaris (muscle) (→ 239)

10 Flexor carpi ulnaris (muscle) (→ 240)

11 Supinator (muscle) (→ 241)
The arrow indicates the direction of pull of the muscle!

12 Flexor digitorum profundus (muscle) (→ 240)

13 Abductor pollicis longus (muscle) (→ 239)

14 Pronator teres (muscle) (→ 241)
Its direction of pull is exactly opposite to the direction of pull of the supinator (muscle)!

15 Extensor pollicis longus (muscle) (→ 241)

16 Extensor pollicis brevis (muscle) (→ 239)

17 Extensor indicis (muscle) (→ 241)

Dorsal view of left arm

The schematic depictions of the muscle origins and insertions are intended to serve as a learning aid to help visualize graphically the origins and insertions of the individual muscles. The **flexor muscles** of the forearm originate rather from the **ulnar and volar**, the **extensor muscles** rather from the **radial and dorsal** side.

11.4.3 Muscle origins and Insertions of Forearm (Ventral View)

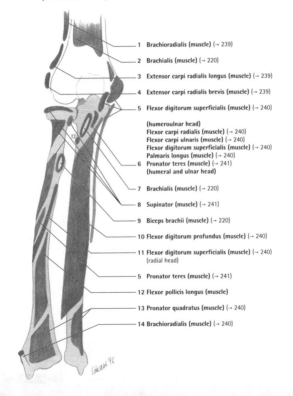

1 Brachioradialis (muscle) (→ 239)

2 Brachialis (muscle) (→ 220)

3 Extensor carpi radialis longus (muscle) (→ 239)

4 Extensor carpi radialis brevis (muscle) (→ 239)

5 Flexor digitorum superficialis (muscle) (→ 240)

(humeroulnar head)
Flexor carpi radialis (muscle) (→ 240)
Flexor carpi ulnaris (muscle) (→ 240)
Flexor digitorum superficialis (muscle) (→ 240)
Palmaris longus (muscle) (→ 240)
6 Pronator teres (muscle) (→ 241)
(humeral and ulnar head)

7 Brachialis (muscle) (→ 220)

8 Supinator (muscle) (→ 241)

9 Biceps brachii (muscle) (→ 220)

10 Flexor digitorum profundus (muscle) (→ 240)

11 Flexor digitorum superficialis (muscle) (→ 240)
(radial head)

5 Pronator teres (muscle) (→ 241)

12 Flexor pollicis longus (muscle)

13 Pronator quadratus (muscle) (→ 240)

14 Brachioradialis (muscle) (→ 240)

11.4.4 Muscles of Forearm (Dorsal View)

1 Brachioradialis (muscle)
IN: Radial nerve
O: Lateral margin of humerus and intermuscular septum
I: Styloid process of radius
F: Flexion at the elbow joint; depending on position, support of supination or pronation

2 Extensor carpi radialis longus (muscle)
IN: Radial nerve
O: Lateral margin and epicondyle of humerus
I: Base of 2nd metacarpal bone
F: Same as 1 and dorsal flexion/abduction at the wrist joint

3 Extensor digitorum (muscle)
IN: Radial nerve
O: Lateral epicondyle
I: Dorsal aponeurosis of digits II to V
F: Ext. at the elbow, wrist/metacarpophalangeal joints

4 Anconeus (muscle)
IN: Radial nerve
O: Lateral epicondyle of humerus
I: Posterior surface of ulna
F: Extension at the elbow joint

5 Extensor carpi ulnaris (muscle)
IN: Radial nerve
O: Lateral epicondyle of humerus
I: Base of 5th metacarpal bone
F: Extension at elbow and wrist joint, abduction towards the ulna at the wrist joint

6 Extensor carpi radialis brevis (muscle)
IN: Radial nerve
O: Lateral epicondyle of humerus
I: Base of 3rd metacarpal bone
F: Same as 1 and dorsal flexion/abduction at the wrist joint

7 Flexor carpi ulnaris (muscle) (→ 240)

8 Extensor digiti minimi (muscle)
IN: Radial nerve
O: Lateral epicondyle of humerus
I: Dorsal aponeurosis of the 5th digit
F: Extension at the elbow, wrist and interphalangeal joint

9 Abductor pollicis longus (muscle)
IN: Radial nerve
O: Posterior surface of ulna, interosseous membrane
I: Base of 1st metacarpal bone
F: Flexion and abduction at the wrist joint, extension and abduction at the carpometacarpal joint

10 Extensor pollicis brevis (muscle)
IN: Radial nerve
O: Posterior surface of radius, interosseous membrane
I: Base of proximal phalanx of thumb
F: Palmar flexion and abduction at the wrist joint, abduction and extension of thumb

11.4.5 Muscles of Forearm (Flexors), Volar View

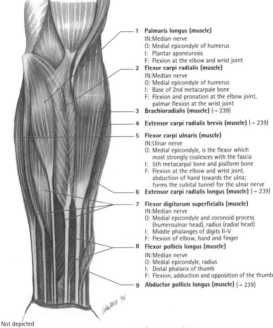

1 Palmaris longus (muscle)
IN: Median nerve
O: Medial epicondyle of humerus
I: Plantar aponeurosis
F: Flexion at the elbow and wrist joint

2 Flexor carpi radialis (muscle)
IN: Median nerve
O: Medial epicondyle of humerus
I: Base of 2nd metacarpale bone
F: Flexion and pronation at the elbow joint,
palmar flexion at the wrist joint

3 Brachioradialis (muscle) (→ 239)

4 Extensor carpi radialis brevis (muscle) (→ 239)

5 Flexor carpi ulnaris (muscle)
IN: Ulnar nerve
O: Medial epicondyle, is the flexor which
most strongly coalesces with the fascia
I: 5th metacarpal bone and pisiform bone
F: Flexion at the elbow and wrist joint,
abduction of hand towards the ulna;
forms the cubital tunnel for the ulnar nerve

6 Extensor carpi radialis longus (muscle) (→ 239)

7 Flexor digitorum superficialis (muscle)
IN: Median nerve
O: Medial epicondyle and coronoid process
(humeroulnar head), radius (radial head)
I: Middle phalanges of digits II–V
F: Flexion of elbow, hand and finger

8 Flexor pollicis longus (muscle)
IN: Median nerve
O: Medial epicondyle, radius
I: Distal phalanx of thumb
F: Flexion, adduction and opposition of the thumb

9 Abductor pollicis longus (muscle) (→ 239)

Not depicted
10 Flexor digitorum profundus (muscle)
IN: Ulnar nerve (ulna), median nerve (radial)
O: Ulna, interosseous membrane
I: Distal phalanges of digits
F: Flexion

11 Pronator quadratus
IN: Median nerve
O: Ulna (anterior margin)
I: Radius (anterior margin)
F: Pronation

11.4.6 Muscles of Arm / Pronator and Supinator

1 Supinator (muscle)
IN: Radial nerve, pierces this muscle
O: Lateral epicondyle, crest of ulna
I: Radius
F: Supination

2 Pronator teres (muscle)
IN: Median nerve, pierces this muscle
O: Medial epicondyle, ulna
I: Lateral and dorsal surface of radius
F: Pronation

Additional deep dorsal muscles (not depicted):

3 Extensor indicis (muscle)s
IN: Radial nerve
O: Posterior surface of ulna
I: Dorsal aponeurosis of 2nd digit
F: Extension at wrist and finger joints

4 Extensor pollicis longus (muscle)
IN: Radial nerve
O: Posterior surface of ulna
I: Distal phalanx of thumb
F: Extension at wrist joint, extension and adduction of the thumb

"SOUPination":   POUR-nation

Supination is to turn your arm palm up, as if you are holding a bowl of soup.

Pronation is to turn your arm with the palm down, as if you are pouring out whatever is in your bowl.

11.4.7 Median, Radial and Ulnar Nerve

The following is a summary of the muscles innervated by the **median, radial and ulnar nerve**.
In addition the consequences of paresis (=partial paralysis) are listed for each individual nerve.

Median nerve		(→ 220)
Innervated muscles	- Flexores carpi radialis, digitorum superficialis, digitorum profundus (radial), pollicis longus and pollicis brevis	(→ 240)
	- Pronatores teres and quadratus	
	- Palmares longus and brevis	(→ 241)
	- Opponens pollicis	(→ 240)
	- Abductor pollicis brevis	(→ 247)
	- Lumbricales (I and/or II)	(→ 246)
		(→ 246)
Paresis	The characteristic clinical sign of median nerve paresis is the **"hand of benediction"**. This occurs because the flexores digitorum superficialis and profundus muscles do not receive any innervation (the ring and little finger can be flexed since the ulnar part of the flexor digitorum profundus (muscle) is supplied by the ulnar nerve).	

Ulnar nerve		(→ 220)
Innervated muscles	- Flexores carpi ulnaris, digitorum profundus (ulnar), digiti minimi brevis and pollicis brevis	(→ 240)
	- Opponentes digiti minimi and pollicis	(→ 247)
	- Abductor digiti minimi	(→ 246)
	- Adductor pollicis	(→ 247)
	- Lumbricales (III and V)	(→ 246)
	- Interossei dorsales and palmares	(→ 248)
Paresis	The characteristic clinical sign of ulnar nerve paresis is the **claw hand**. It results from the failure of the lumbricales and interossei muscles. These would normally flex at the metacarpophalangeal joints and extend at the two distal interphalangeal joints. In paresis the exact opposite occurs: Flexion of the distal and medial phalanges and at the same time extension at the metacarpophalangeal joints.	

Radial nerve		(→ 219)
Innervated muscles	- All extensors of the forearm!!	(→ 239)
	- Supinator	(→ 241)
	- Brachioradialis	(→ 239)
	- Anconeus	(→ 239)
	- Triceps brachii	(→ 232)
Paresis	The characteristic clinical sign of radial nerve paresis is **wrist-drop**. Neither the wrist joint nor the interphalangeal joints can be extended during radial nerve paresis, since the radial nerve innervates all extensors of the forearm. The wrist joint remains in a position of palmar flexion.	

11.5 Hand

11.5.1 Ligaments of Hand (Dorsal View)

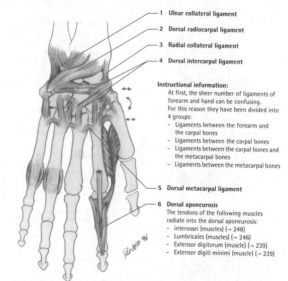

1 Ulnar collateral ligament

2 Dorsal radiocarpal ligament

3 Radial collateral ligament

4 Dorsal intercarpal ligament

Instructional information:
At first, the sheer number of ligaments of forearm and hand can be confusing.
For this reason they have been divided into 4 groups:
- Ligaments between the forearm and the carpal bones
- Ligaments between the carpal bones
- Ligaments between the carpal bones and the metacarpal bones
- Ligaments between the metacarpal bones

5 Dorsal metacarpal ligament

6 Dorsal aponeurosis
The tendons of the following muscles radiate into the dorsal aponeurosis:
- interossei (muscles) (→ 248)
- Lumbricales (muscles) (→ 246)
- Extensor digitorum (muscle) (→ 239)
- Extensor digiti minimi (muscle) (→ 239)

Dorsal view of hand

Mnemonic of the carpal bones:

"Some Lovers Try Positions That They Can't Handle."

(Scaphoid, Lunate, Triquetrum, Pisiform, Trapezium, Trapezoid, Capitate and Hamate)

11.5.2 Ligaments of Hand, Course of Tendons

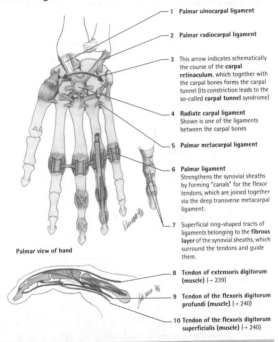

1 Palmar ulnocarpal ligament

2 Palmar radiocarpal ligament

3 This arrow indicates schematically the course of the **carpal retinaculum**, which together with the carpal bones forms the carpal tunnel (its constriction leads to the so-called **carpal tunnel** syndrome)

4 **Radiate carpal ligament**
Shown is one of the ligaments between the carpal bones

5 **Palmar metacarpal ligament**

6 **Palmar ligament**
Strengthens the synovial sheaths by forming "canals" for the flexor tendons, which are joined together via the deep transverse metacarpal ligament.

7 Superficial ring-shaped tracts of ligaments belonging to the **fibrous layer** of the synovial sheaths, which surround the tendons and guide them.

Palmar view of hand

8 **Tendon of extensoris digitorum (muscle)** (→ 239)

9 **Tendon of the flexoris digitorum profundi (muscle)** (→ 240)

10 **Tendon of the flexoris digitorum superficialis (muscle)** (→ 240)

The tendon of the flexor digitorum profundus (muscle) (→ 240) **pierces** the tendon of the flexor digitorum superficialis (muscle) (→ 240) from below, at the level of the proximal phalanx. It also gives rise to the lumbricales (muscles), which then project to the dorsal aponeurosis. The flexor tendons are fixed to the phalanges via short ligaments.

11.5.3 Muscles and Tendon Insertions of Hand (Dorsal View)

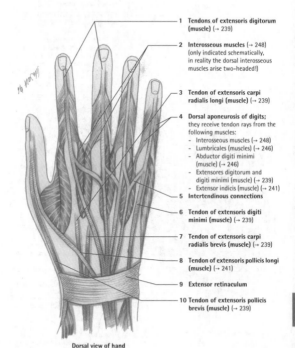

1 Tendons of extensoris digitorum (muscle) (→ 239)

2 Interosseous muscles (→ 248) (only indicated schematically, in reality the dorsal interosseous muscles arise two-headed!)

3 Tendon of extensoris carpi radialis longi (muscle) (→ 239)

4 Dorsal aponeurosis of digits; they receive tendon rays from the following muscles:
   - Interosseous muscles (→ 248)
   - Lumbricales (muscles) (→ 246)
   - Abductor digiti minimi (muscle) (→ 246)
   - Extensores digitorum and digiti minimi (muscle) (→ 239)
   - Extensor indicis (muscle) (→ 241)

5 Intertendinous connections

6 Tendon of extensoris digiti minimi (muscle) (→ 239)

7 Tendon of extensoris carpi radialis brevis (muscle) (→ 239)

8 Tendon of extensoris pollicis longi (muscle) (→ 241)

9 Extensor retinaculum

10 Tendon of extensoris pollicis brevis (muscle) (→ 239)

Dorsal view of hand

11.5.4 Short Flexors of Hand (Superficial Layer)

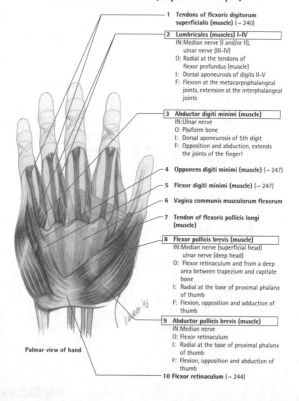

1 **Tendons of flexoris digitorum superficialis (muscle)** (→ 240)

2 **Lumbricales (muscles) I-IV**
IN: Median nerve (I and/or II),
    ulnar nerve (III-IV)
O: Radial at the tendons of
   flexor profundus (muscle)
I: Dorsal aponeurosis of digits II-V
F: Flexion at the metacarpophalangeal
   joints, extension at the interphalangeal
   joints

3 **Abductor digiti minimi (muscle)**
IN: Ulnar nerve
O: Pisiform bone
I: Dorsal aponeurosis of 5th digit
F: Opposition and abduction, extends
   the joints of the finger!

4 **Opponens digiti minimi (muscle)** (→ 247)

5 **Flexor digiti minimi (muscle)** (→ 247)

6 **Vagina communis musculorum flexorum**

7 **Tendon of flexoris pollicis longi
   (muscle)**

8 **Flexor pollicis brevis (muscle)**
IN: Median nerve (superficial head)
    ulnar nerve (deep head)
O: Flexor retinaculum and from a deep
   area between trapezium and capitate
   bone
I: Radial at the base of proximal phalanx
   of thumb
F: Flexion, opposition and adduction of
   thumb

9 **Abductor pollicis brevis (muscle)**
IN: Median nerve
O: Flexor retinaculum
I: Radial at the base of proximal phalanx
   of thumb
F: Flexion, opposition and abduction of
   thumb

Palmar view of hand

10 **Flexor retinaculum** (→ 244)

11.5.5 Short Muscles of Hand (Deep Layer)

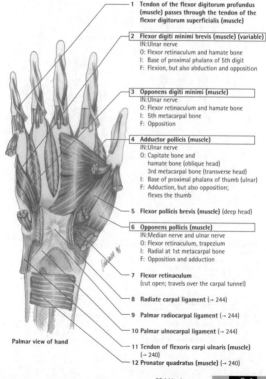

1 **Tendon of the flexor digitorum profundus (muscle) passes through the tendon of the flexor digitorum superficialis (muscle)**

2 **Flexor digiti minimi brevis (muscle) (variable)**
IN: Ulnar nerve
O: Flexor retinaculum and hamate bone
I: Base of proximal phalanx of 5th digit
F: Flexion, but also abduction and opposition

3 **Opponens digiti minimi (muscle)**
IN: Ulnar nerve
O: Flexor retinaculum and hamate bone
I: 5th metacarpal bone
F: Opposition

4 **Adductor pollicis (muscle)**
IN: Ulnar nerve
O: Capitate bone and
   hamate bone (oblique head)
   3rd metacarpal bone (transverse head)
I: Base of proximal phalanx of thumb (ulnar)
F: Adduction, but also opposition;
   flexes the thumb

5 **Flexor pollicis brevis (muscle)** (deep head)

6 **Opponens pollicis (muscle)**
IN: Median nerve and ulnar nerve
O: Flexor retinaculum, trapezium
I: Radial at 1st metacarpal bone
F: Opposition and adduction

7 **Flexor retinaculum**
(cut open; travels over the carpal tunnel)

8 **Radiate carpal ligament** (→ 244)

9 **Palmar radiocarpal ligament** (→ 244)

10 **Palmar ulnocarpal ligament** (→ 244)

11 **Tendon of flexoris carpi ulnaris (muscle)**
(→ 240)

12 **Pronator quadratus (muscle)** (→ 240)

Palmar view of hand

11.5.6 Interosseous Muscles

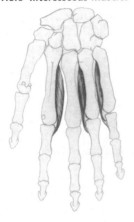 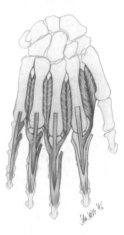

Palmar interosseous muscles   **Dorsal interosseous muscles**

Palmar interosseous muscles (are oriented away from the middle finger!!!)
IN: Ulnar nerve
O: One-headed at the metacarpal bones 2 through 5
I: Dorsal aponeurosis of digits II–V
F: Flexion and adduction at the metacarpophalangeal joints II–IV;
    extension of the interphalangeal joints II–IV

Dorsal interosseous muscles (are oriented towards the middle finger!!!)
IN: Ulnar nerve
O: Two-headed at adjacent sides of two metacarpals of metacarpal bone I–IV
I: Dorsal aponeurosis of digits II–IV
F: Flexion and abduction at the metacarpophalangeal joints II–IV;
    extension of the interphalangeal joints II–V

11.5.7 Muscle Origins and Insertions of Hand (Palmar View)

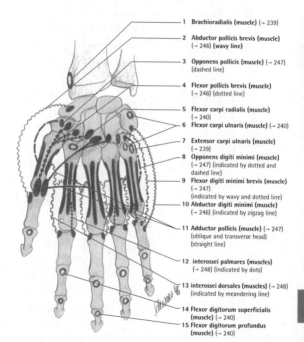

1 Brachioradialis (muscle) (→ 239)

2 Abductor pollicis brevis (muscle) (→ 246) (wavy line)

3 Opponens pollicis (muscle) (→ 247) (dashed line)

4 Flexor pollicis brevis (muscle) (→ 246) (dotted line)

5 Flexor carpi radialis (muscle) (→ 240)

6 Flexor carpi ulnaris (muscle) (→ 240)

7 Extensor carpi ulnaris (muscle) (→ 239)

8 Opponens digiti minimi (muscle) (→ 247) (indicated by dotted and dashed line)

9 Flexor digiti minimi brevis (muscle) (→ 247) (indicated by wavy and dotted line)

10 Abductor digiti minimi (muscle) (→ 246) (indicated by zigzag line)

11 Adductor pollicis (muscle) (→ 247) (oblique and transverse head) (straight line)

12 interossei palmares (muscles) (→ 248) (indicated by dots)

13 interossei dorsales (muscles) (→ 248) (indicated by meandering line)

14 Flexor digitorum superficialis (muscle) (→ 240)

15 Flexor digitorum profundus (muscle) (→ 240)

11.5.8 Arteries of Hand

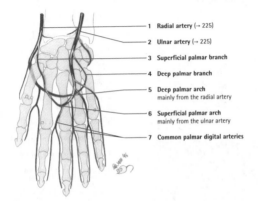

1 Radial artery (→ 225)

2 Ulnar artery (→ 225)

3 Superficial palmar branch

4 Deep palmar branch

5 Deep palmar arch
   mainly from the radial artery

6 Superficial palmar arch
   mainly from the ulnar artery

7 Common palmar digital arteries

11.6 Nerves of Skin, Dermatomes
11.6.1 Nerves of Skin and Dermatomes of Arm

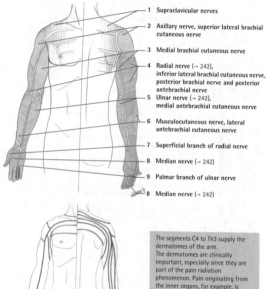

1 Supraclavicular nerves

2 Axillary nerve, superior lateral brachial cutaneous nerve

3 Medial brachial cutaneous nerve

4 Radial nerve (→ 242),
inferior lateral brachial cutaneous nerve,
posterior brachial nerve and posterior antebrachial nerve

5 Ulnar nerve (→ 242),
medial antebrachial cutaneous nerve

6 Musculocutaneous nerve, lateral antebrachial cutaneous nerve

7 Superficial branch of radial nerve

8 Median nerve (→ 242)

9 Palmar branch of ulnar nerve

8 Median nerve (→ 242)

The segments C4 to Th3 supply the dermatomes of the arm.
The dermatomes are clinically important, especially since they are part of the pain radiation phenomenon. Pain originating from the inner organs, for example, is projected to corresponding dermatomes of the body surface (Head's zones).

11.6.2 Nerves of Skin

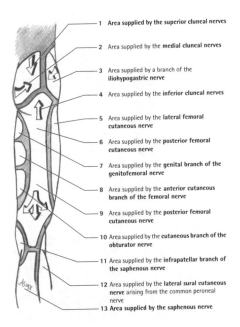

1 Area supplied by the **superior cluneal nerves**

2 Area supplied by the **medial cluneal nerves**

3 Area supplied by a branch of the **iliohypogastric nerve**

4 Area supplied by the **inferior cluneal nerves**

5 Area supplied by the **lateral femoral cutaneous nerve**

6 Area supplied by the **posterior femoral cutaneous nerve**

7 Area supplied by the **genital branch of the genitofemoral nerve**

8 Area supplied by the **anterior cutaneous branch of the femoral nerve**

9 Area supplied by the **posterior femoral cutaneous nerve**

10 Area supplied by the **cutaneous branch of the obturator nerve**

11 Area supplied by the **infrapatellar branch of the saphenous nerve**

12 Area supplied by the **lateral sural cutaneous nerve** arising from the common peroneal nerve

13 Area supplied by the **saphenous nerve**

12. Lower Limb

12.1 Pelvis

12.1.1 Bones and Ligaments of Pelvis

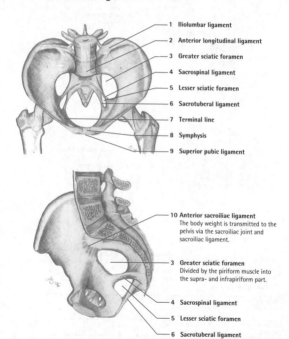

1 Iliolumbar ligament

2 Anterior longitudinal ligament

3 Greater sciatic foramen

4 Sacrospinal ligament

5 Lesser sciatic foramen

6 Sacrotuberal ligament

7 Terminal line

8 Symphysis

9 Superior pubic ligament

10 Anterior sacroiliac ligament
The body weight is transmitted to the pelvis via the sacroiliac joint and sacroiliac ligament.

3 Greater sciatic foramen
Divided by the piriform muscle into the supra- and infrapiriform part.

4 Sacrospinal ligament

5 Lesser sciatic foramen

6 Sacrotuberal ligament

12.1.2 Ligaments of Hip

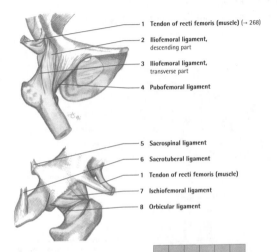

1 Tendon of recti femoris (muscle) (→ 268)
2 Iliofemoral ligament, descending part
3 Iliofemoral ligament, transverse part
4 Pubofemoral ligament

5 Sacrospinal ligament
6 Sacrotuberal ligament
1 Tendon of recti femoris (muscle)
7 Ischiofemoral ligament
8 Orbicular ligament

Ligament	Course of ligament	Adduction	Abduction	Retro-version	Ante-version	Lateral rotation	Medial rotation
Iliofemoral ligament Transverse part	From the anterior inferior iliac spine to the intertrochanteric line	↓		↓		↓	
Descending part			↓	↓			↓
Pubofemoral ligament	From the upper ramus of pubic bone to the intertrochanteric line		↓	↓			↓
Ischiofemoral ligament	From the ischial bone to the trochanteric fossa		↓	↓			↓
Orbicular ligament	Not attached to any bone:	↓: Suppression					

12.1.3 Ligaments of Hip Joint

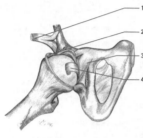

1 **Tendon of recti femoris (muscle)** (→ 268)

2 **Iliofemoral ligament**

3 **Pubofemoral ligament**

4 **Ligament of head of femur**
It lies within the fatty tissue, which occupies the acetabular fossa. It travels from the acetabular incisure and the transverse acetabular ligament to the fovea of femoral head. It carries small arterial branches, important for nourishment of the fermural head in the child, which later obliterate in the adult.

The **hip joint** is a ball-and-socket type joint (that is to say that the acetabulum surrounds the femoral head beyond the equator). It has three degrees of freedom. Only at the lunate fasciae the participating bones have contact with each other.

The hip joint is strongly reinforced and guided by ligaments. The anatomical orientation of the ligaments defines the movement possibilities (anteversion, retroversion, medial rotation, lateral rotation, abduction, adduction).

Clinical Information

Aseptic bone necrosis of the femoral head in juveniles is known as Perthes disease.

12.1.4 Gluteal Region of Male and Female

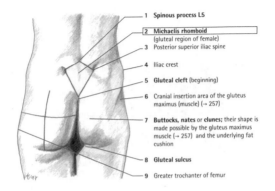

1 Spinous process L5

2 **Michaelis rhomboid**
(gluteal region of female)

3 Posterior superior iliac spine

4 Iliac crest

5 **Gluteal cleft** (beginning)

6 Cranial insertion area of the gluteus
maximus (muscle) (→ 257)

7 **Buttocks, nates** or **clunes**; their shape is
made possible by the gluteus maximus
muscle (→ 257) and the underlying fat
cushion

8 **Gluteal sulcus**

9 Greater trochanter of femur

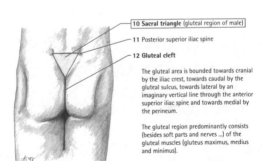

10 **Sacral triangle** (gluteal region of male)

11 Posterior superior iliac spine

12 **Gluteal cleft**

The gluteal area is bounded towards cranial
by the iliac crest, towards caudal by the
gluteal sulcus, towards lateral by an
imaginary vertical line through the anterior
superior iliac spine and towards medial by
the perineum.

The gluteal region predominantly consists
(besides soft parts and nerves ...) of the
gluteal muscles (gluteus maximus, medius
and minimus).

12.1.5 Gluteal Muscles (Dorsal View)

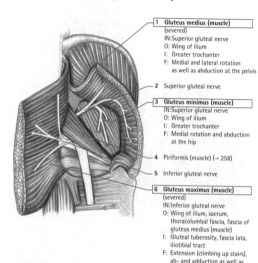

1 Gluteus medius (muscle)
(severed)
IN: Superior gluteal nerve
O: Wing of ilium
I: Greater trochanter
F: Medial and lateral rotation
as well as abduction at the pelvis

2 Superior gluteal nerve

3 Gluteus minimus (muscle)
IN: Superior gluteal nerve
O: Wing of ilium
I: Greater trochanter
F: Medial rotation and abduction
at the hip

4 Piriformis (muscle) (→ 258)

5 Inferior gluteal nerve

6 Gluteus maximus (muscle)
(severed)
IN: Inferior gluteal nerve
O: Wing of ilium, sacrum,
thoracolumbal fascia, fascia of
gluteus medius (muscle)
I: Gluteal tuberosity, fascia lata,
iliotibial tract
F: Extension (climbing up stairs),
ab- and adduction as well as
lateral rotation at the hip joint

Foramen	Traversing structures
Suprapiriform foramen	Superior gluteal artery, vein and nerve
Infrapiriform foramen	inferior gluteal artery, vein and nerve; internal pudendal artery, vein and nerve; sciatic nerve and posterior femoral cutaneous nerve
Lesser sciatic foramen (→ 209)	Internal pudendal artery, vein and nerve (→ 285); at this location they project back into the pelvis and travel within the Alcock's canal (→ 286)

The above figure shows the muscles, which are innervated by the superior and inferior gluteal nerve.

12.1.6 Muscles of Gluteal Region (Dorsal View)

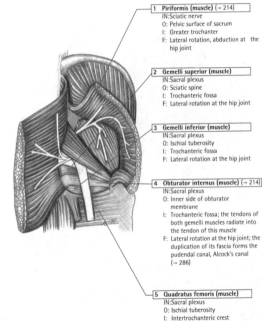

1 Piriformis (muscle) (→ 214)
IN: Sciatic nerve
O: Pelvic surface of sacrum
I: Greater trochanter
F: Lateral rotation, abduction at the hip joint

2 Gemelli superior (muscle)
IN: Sacral plexus
O: Sciatic spine
I: Trochanteric fossa
F: Lateral rotation at the hip joint

3 Gemelli inferior (muscle)
IN: Sacral plexus
O: Ischial tuberosity
I: Trochanteric fossa
F: Lateral rotation at the hip joint

4 Obturator internus (muscle) (→ 214)
IN: Sacral plexus
O: Inner side of obturator membrane
I: Trochanteric fossa; the tendons of both gemelli muscles radiate into the tendon of this muscle
F: Lateral rotation at the hip joint; the duplication of its fascia forms the pudendal canal, Alcock's canal (→ 286)

5 Quadratus femoris (muscle)
IN: Sacral plexus
O: Ischial tuberosity
I: Intertrochanteric crest
F: Lateral rotation and adduction at the hip joint

12.1.7 Muscle Insertions and Origins of Pelvis

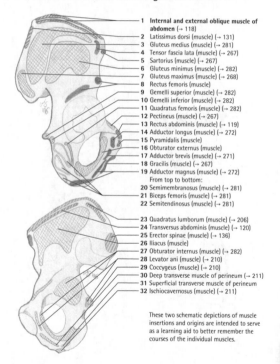

1. **Internal and external oblique muscle of abdomen** (→ 118)
2. Latissimus dorsi (muscle) (→ 131)
3. Gluteus medius (muscle) (→ 281)
4. Tensor fascia lata (muscle) (→ 267)
5. Sartorius (muscle) (→ 267)
6. Gluteus minimus (muscle) (→ 282)
7. Gluteus maximus (muscle) (→ 268)
8. Rectus femoris (muscle)
9. Gemelli superior (muscle) (→ 282)
10. Gemelli inferior (muscle) (→ 282)
11. Quadratus femoris (muscle) (→ 282)
12. Pectineus (muscle) (→ 267)
13. Rectus abdominis (muscle) (→ 119)
14. Adductor longus (muscle) (→ 272)
15. Pyramidalis (muscle)
16. Obturator externus (muscle)
17. Adductor brevis (muscle) (→ 271)
18. Gracilis (muscle) (→ 267)
19. Adductor magnus (muscle) (→ 272)
    From top to bottom:
20. Semimembranosus (muscle) (→ 281)
21. Biceps femoris (muscle) (→ 281)
22. Semitendinosus (muscle) (→ 281)

23. Quadratus lumborum (muscle) (→ 206)
24. Transversus abdominis (muscle) (→ 120)
25. Erector spinae (muscle) (→ 136)
26. Iliacus (muscle)
27. Obturator internus (muscle) (→ 282)
28. Levator ani (muscle) (→ 210)
29. Coccygeus (muscle) (→ 210)
30. Deep transverse muscle of perineum (→ 211)
31. Superficial transverse muscle of perineum
32. Ischiocavernosus (muscle) (→ 211)

These two schematic depictions of muscle insertions and origins are intended to serve as a learning aid to better remember the courses of the individual muscles.

12.1.8 Trendelenburg's Sign

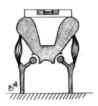

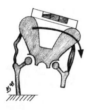

Clinical Information

The gluteal muscles fix the position of the pelvis relative to the weight bearing leg (the leg that carries the weight of the body). In this example, we assume that the **right** gluteal muscles have been injured. As long as the body weight rests on both legs, no action by the gluteal muscles is required to maintain the pelvis in a horizontal position. Under these circumstances, the functional deficiency of the right gluteal muscle group does not manifest itself.

In a situation where the right leg is weight bearing and the left is non-weight bearing, the right **gluteal muscles** hold the pelvis in a horizontal position. If in this situation the right gluteal muscles are **weakened or paralyzed,** then pelvic fixation cannot be maintained and tilting toward the non-affected side results. Clinicians call this the **positive Trendelenburg's sign.** This finding can also be observed in luxations of the hip joint.

12.1.9 Intramuscular Injection (I.M.)

1 **Eminence of iliac crest**

2 Anterior superior iliac spine

3 Greater trochanter

Clinical Information

The **method according to Hochstetter** requires the following procedure for finding the correct I.M. injection site and for preventing damage to nerves and vessels:

The patient lies on the side with legs adducted. Use the index- and middle finger to find the anterior superior iliac spine and the eminence of iliac crest. The heel of hand must be placed on the greater trochanter. To accomplish this, move the hand approximately 2 cm in the direction indicated by the figure, while keeping the angle between index- and middle finger unchanged. During this process keep the middle finger placed on the anterior superior iliac spine. The resulting region between the straddled fingers represents the area in which injection should occur.

The needle should not point directly towards the body axis but instead should be oriented slightly towards ventral and cranial. To ensure no blood vessel has been punctured, aspirate for control purposes. No trace of blood should be seen in the syringe. Clinical studies indicate that, instead of the gluteus medius muscle, a high percentage of I.M. injections are erroneously instilled into the subcutaneous fatty tissue. Therefore it is important to insert the needle sufficiently deep.

An alternative procedure is the **crista method according to Sachtleben**. The injection site lies 3 finger diameters below the iliac crest on an imaginary line between the greater trochanter and the eminence of iliac crest.

12.1.10 Arteries of Leg

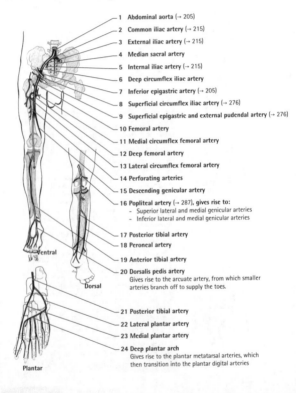

1 Abdominal aorta (→ 205)

2 Common iliac artery (→ 215)

3 External iliac artery (→ 215)

4 Median sacral artery

5 Internal iliac artery (→ 215)

6 Deep circumflex iliac artery

7 Inferior epigastric artery (→ 205)

8 Superficial circumflex iliac artery (→ 276)

9 Superficial epigastric and external pudendal artery (→ 276)

10 Femoral artery

11 Medial circumflex femoral artery

12 Deep femoral artery

13 Lateral circumflex femoral artery

14 Perforating arteries

15 Descending genicular artery

16 Popliteal artery (→ 287), gives rise to:
- Superior lateral and medial genicular arteries
- Inferior lateral and medial genicular arteries

17 Posterior tibial artery

18 Peroneal artery

19 Anterior tibial artery

20 Dorsalis pedis artery
Gives rise to the arcuate artery, from which smaller arteries branch off to supply the toes.

21 Posterior tibial artery

22 Lateral plantar artery

23 Medial plantar artery

24 Deep plantar arch
Gives rise to the plantar metatarsal arteries, which then transition into the plantar digital arteries

Ventral

Dorsal

Plantar

12.1.11 Superficial Veins of Leg

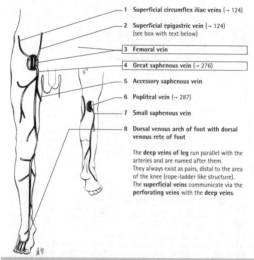

1 Superficial circumflex iliac veins (→ 124)

2 Superficial epigastric vein (→ 124)
(see box with text below)

3 Femoral vein

4 Great saphenous vein (→ 276)

5 Accessory saphenous vein

6 Popliteal vein (→ 287)

7 Small saphenous vein

8 Dorsal venous arch of foot with dorsal venous rete of foot

The **deep veins of leg** run parallel with the arteries and are named after them.
They always exist as pairs, distal to the area of the knee (rope-ladder like structure).
The **superficial veins** communicate via the **perforating veins** with the **deep veins**.

At this point the **portocaval anastomoses** should again be called into remembrance. These are alternative routes of blood supply, that originate from the portal vein of the liver and end in the superior and inferior vena cava. They bypass the liver during impairment of portal flow. The following routes are available:
- via the veins of the lesser curvature of the stomach and the esophagus to the azygous vein and then to the superior vena cava.
- via the rectal veins and the rectal venous plexus to the internal iliac vein and then to the inferior vena cava.
- via the paraumbilical veins to veins of the skin, for example to the superficial epigastric vein (looks very impressive and is also known as "Medusa's head") and then via the femoral vein to the inferior vena cava.

Clinical Information
Abnormal dilatations of the superficial venous system are called **varixes** (varicose veins). This pathologic condition results from incompetent venous valves, which impair normal drainage. The great saphenous vein is frequently used in bypass operations for circumventing incompetent coronary vessels.

12.1.12 Nerves of Leg

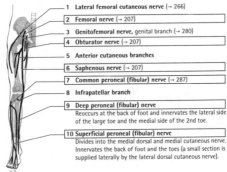

1 **Lateral femoral cutaneous nerve** (→ 266)

2 **Femoral nerve** (→ 207)

3 **Genitofemoral nerve**, genital branch (→ 280)

4 **Obturator nerve** (→ 207)

5 **Anterior cutaneous branches**

6 **Saphenous nerve** (→ 207)

7 **Common peroneal (fibular) nerve** (→ 287)

8 **Infrapatellar branch**

9 **Deep peroneal (fibular) nerve**
Reoccurs at the back of foot and innervates the lateral side of the large toe and the medial side of the 2nd toe.

10 **Superficial peroneal (fibular) nerve**
Divides into the medial dorsal and medial cutaneous nerve. Innervates the back of foot and the toes (a small section is supplied laterally by the lateral dorsal cutaneous nerve).

Ventral

11 **Superior and inferior gluteal nerves**

12 **Posterior femoral cutaneous nerve**

13 **Sciatic nerve** (→ 285)

14 **Common peroneal (fibular) nerve**
It loops below the fibular head towards anterior and divides into the deep and superficial peroneal nerves. It also gives off the lateral sural cutaneous nerve.

15 **Tibial nerve** (→ 287)
More downstream it gives off the medial sural cutaneous nerve, which merges with the lateral sural cutaneous nerve to form the sural nerve. Later this nerve becomes the lateral dorsal cutaneous nerve.
The tibial nerve projects, posterior to the medial malleolus, downward and divides into the medial and lateral plantar nerves.

16 **Lateral plantar nerve**

17 **Medial plantar nerve**

Dorsal

12.1.13 Lumbosacral Plexus

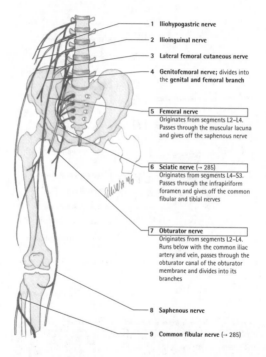

1 **Iliohypogastric nerve**

2 **Ilioinguinal nerve**

3 **Lateral femoral cutaneous nerve**

4 **Genitofemoral nerve;** divides into the **genital and femoral branch**

5 **Femoral nerve**
Originates from segments L2–L4. Passes through the muscular lacuna and gives off the saphenous nerve

6 **Sciatic nerve** (→ 285)
Originates from segments L4–S3. Passes through the infrapiriform foramen and gives off the common fibular and tibial nerves

7 **Obturator nerve**
Originates from segments L2–L4. Runs below with the common iliac artery and vein, passes through the obturator canal of the obturator membrane and divides into its branches

8 **Saphenous nerve**

9 **Common fibular nerve** (→ 285)

12.1.14 Schematic of Lumbosacral Plexus

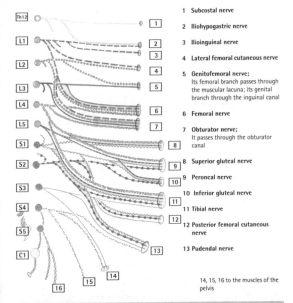

1 Subcostal nerve

2 Iliohypogastric nerve

3 Ilioinguinal nerve

4 Lateral femoral cutaneous nerve

5 Genitofemoral nerve;
Its femoral branch passes through
the muscular lacuna; its genital
branch through the inguinal canal

6 Femoral nerve

7 Obturator nerve;
It passes through the obturator
canal

8 Superior gluteal nerve

9 Peroneal nerve

10 Inferior gluteal nerve

11 Tibial nerve

12 Posterior femoral cutaneous
nerve

13 Pudendal nerve

14, 15, 16 to the muscles of the
pelvis

The nerves numbered 2–7 are very important and should therefore be memorized.
The following mnemonic can be derived from the initial letters of these nerves:

"Interested In Getting Laid On Fridays?"

Some authors subdivide the lumbosacral plexus (→ 207) into the lumbar and sacral plexus.
The above mnemonic applies to the nerves of the lumbar plexus.

12.2 Anterior Femoral Region

12.2.1 Muscles of Thigh (Ventral View)

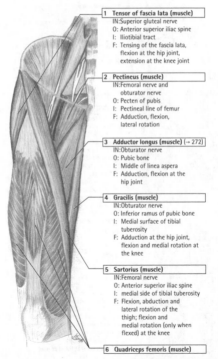

1 Tensor of fascia lata (muscle)
IN: Superior gluteal nerve
O: Anterior superior iliac spine
I: Iliotibial tract
F: Tensing of the fascia lata,
  flexion at the hip joint,
  extension at the knee joint

2 Pectineus (muscle)
IN: Femoral nerve and
  obturator nerve
O: Pecten of pubis
I: Pectineal line of femur
F: Adduction, flexion,
  lateral rotation

3 Adductor longus (muscle) (→ 272)
IN: Obturator nerve
O: Pubic bone
I: Middle of linea aspera
F: Adduction, flexion at the
  hip joint

4 Gracilis (muscle)
IN: Obturator nerve
O: Inferior ramus of pubic bone
I: Medial surface of tibial
  tuberosity
F: Adduction at the hip joint,
  flexion and medial rotation at
  the knee

5 Sartorius (muscle)
IN: Femoral nerve
O: Anterior superior iliac spine
I: medial side of tibial tuberosity
F: Flexion, abduction and
  lateral rotation of the
  thigh; flexion and
  medial rotation (only when
  flexed) at the knee

6 Quadriceps femoris (muscle)

12.2.2 Muscles of Thigh (Lateral View)

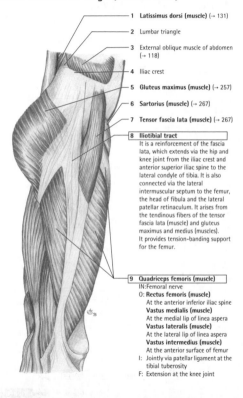

1 **Latissimus dorsi (muscle)** (→ 131)

2 Lumbar triangle

3 External oblique muscle of abdomen (→ 118)

4 Iliac crest

5 **Gluteus maximus (muscle)** (→ 257)

6 **Sartorius (muscle)** (→ 267)

7 **Tensor fascia lata (muscle)** (→ 267)

8 **Iliotibial tract**
It is a reinforcement of the fascia lata, which extends via the hip and knee joint from the iliac crest and anterior superior iliac spine to the lateral condyle of tibia. It is also connected via the lateral intermuscular septum to the femur, the head of fibula and the lateral patellar retinaculum. It arises from the tendinous fibers of the tensor fascia lata (muscle) and gluteus maximus and medius (muscles). It provides tension-banding support for the femur.

9 **Quadriceps femoris (muscle)**
IN: Femoral nerve
O: **Rectus femoris (muscle)**
   At the anterior inferior iliac spine
  Vastus medialis (muscle)
   At the medial lip of linea aspera
  Vastus lateralis (muscle)
   At the lateral lip of linea aspera
  Vastus intermedius (muscle)
   At the anterior surface of femur
I: Jointly via patellar ligament at the tibial tuberosity
F: Extension at the knee joint

12.2.3 Muscular and Vascular Lacunae

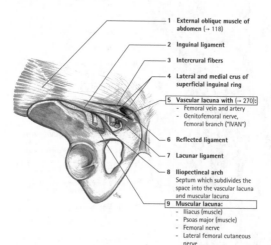

1 **External oblique muscle of abdomen** (→ 118)

2 **Inguinal ligament**

3 **Intercrural fibers**

4 **Lateral and medial crus of superficial inguinal ring**

5 **Vascular lacuna with** (→ 270):
   - Femoral vein and artery
   - Genitofemoral nerve, femoral branch ("IVAN")

6 **Reflected ligament**

7 **Lacunar ligament**

8 **Iliopectineal arch**
   Septum which subdivides the space into the vascular lacuna and muscular lacuna

9 **Muscular lacuna:**
   - Iliacus (muscle)
   - Psoas major (muscle)
   - Femoral nerve
   - Lateral femoral cutaneous nerve

Important structures traversing the lacunae:
Vascular lacuna (medial):
- Femoral vein and artery - Genitofemoral nerve, femoral branch - Lymphatic ducts
Muscular lacuna (lateral):
- Iliopsoas (muscle) (= Iliacus and psoas major (muscle)) - Femoral nerve - Lateral femoral cutaneous nerve

12.2.4 Anterior Nerves and Vessels of Thigh

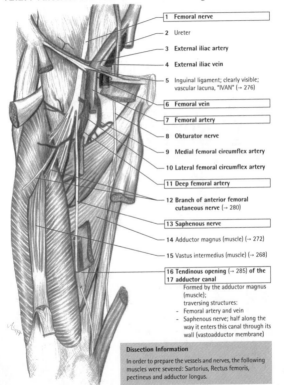

1 Femoral nerve

2 Ureter

3 External iliac artery

4 External iliac vein

5 Inguinal ligament; clearly visible; vascular lacuna, "IVAN" (→ 276)

6 Femoral vein

7 Femoral artery

8 Obturator nerve

9 Medial femoral circumflex artery

10 Lateral femoral circumflex artery

11 Deep femoral artery

12 Branch of anterior femoral cutaneous nerve (→ 280)

13 Saphenous nerve

14 Adductor magnus (muscle) (→ 272)

15 Vastus intermedius (muscle) (→ 268)

16 Tendinous opening (→ 285) of the
17 adductor canal
Formed by the adductor magnus (muscle);
traversing structures:
- Femoral artery and vein
- Saphenous nerve; half along the way it enters this canal through its wall (vastoadductor membrane)

Dissection Information

In order to prepare the vessels and nerves, the following muscles were severed: Sartorius, Rectus femoris, pectineus and adductor longus.

12.2.5 Adductors of Thigh I

1 **Adductor minimus (muscle)**

2 **Adductor brevis (muscle)**

3 **Adductor longus (muscle)** (→ 271)

4 **Adductor magnus (muscle)** (→ 271)
Clearly visible is the gap in the line of attachment (shown here as black areas!). This gap represents the area traversed by the adductor canal.

Attention: This schematic shows the muscle attachments of the femur from **posterior**!

5 **Adductor brevis (muscle)**
IN: Obturator nerve
O: Pubic bone, inferior ramus
I: Medial lip of linea aspera, relative proximal
F: Adduction of leg, extension and lateral rotation at the hip joint, stabilization of pelvis when standing and during locomotion

6 **Adductor magnus (muscle), adductor canal**
(→ 271)
The adductor longus muscle would lie ventral and slightly inferior to the adductor brevis muscle; the adductor magnus muscle runs more distal than the adductor longus muscle and slightly dorsal to this muscle.
The adductor muscles lie **within the inguinal region** as follows: The adductor magnus muscle runs most dorsally; the remaining adductors travel from the inner side of the thigh towards the femur according to the following sequence:
Adductor longus muscle, adductor brevis muscle, adductor minimus muscle

12.2.6 Adductors of Thigh II

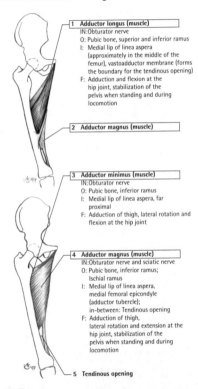

1 Adductor longus (muscle)
IN: Obturator nerve
O: Pubic bone, superior and inferior ramus
I: Medial lip of linea aspera
(approximately in the middle of the
femur), vastoadductor membrane (forms
the boundary for the tendinous opening)
F: Adduction and flexion at the
hip joint, stabilization of the
pelvis when standing and during
locomotion

2 Adductor magnus (muscle)

3 Adductor minimus (muscle)
IN: Obturator nerve
O: Pubic bone, inferior ramus
I: Medial lip of linea aspera, far
proximal
F: Adduction of thigh, lateral rotation and
flexion at the hip joint

4 Adductor magnus (muscle)
IN: Obturator nerve and sciatic nerve
O: Pubic bone, inferior ramus;
Ischial ramus
I: Medial lip of linea aspera,
medial femoral epicondyle
(adductor tubercle);
in-between: Tendinous opening
F: Adduction of thigh,
lateral rotation and extension at the
hip joint, stabilization of the
pelvis when standing and during
locomotion

5 Tendinous opening

12.2.7 Muscle Origins and Insertions of Thigh

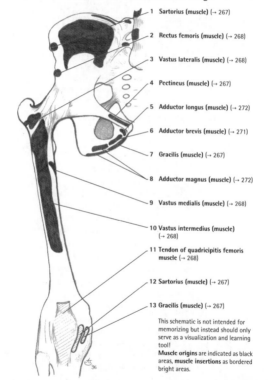

1 Sartorius (muscle) (→ 267)

2 Rectus femoris (muscle) (→ 268)

3 Vastus lateralis (muscle) (→ 268)

4 Pectineus (muscle) (→ 267)

5 Adductor longus (muscle) (→ 272)

6 Adductor brevis (muscle) (→ 271)

7 Gracilis (muscle) (→ 267)

8 Adductor magnus (muscle) (→ 272)

9 Vastus medialis (muscle) (→ 268)

10 Vastus intermedius (muscle) (→ 268)

11 Tendon of quadricipitis femoris muscle (→ 268)

12 Sartorius (muscle) (→ 267)

13 Gracilis (muscle) (→ 267)

This schematic is not intended for memorizing but instead should only serve as a visualization and learning tool!
Muscle origins are indicated as black areas, **muscle insertions** as bordered bright areas.

12.2.8 Muscle Origins and Insertions of Femur

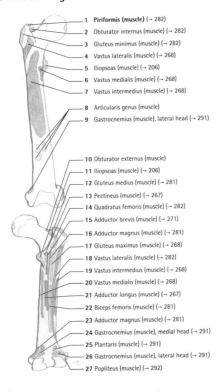

1 **Piriformis (muscle)** (→ 282)
2 Obturator internus (muscle) (→ 282)
3 Gluteus minimus (muscle) (→ 282)
4 Vastus lateralis (muscle) (→ 268)
5 Iliopsoas (muscle) (→ 206)
6 Vastus medialis (muscle) (→ 268)
7 Vastus intermedius (muscle) (→ 268)
8 Articularis genus (muscle)
9 Gastrocnemius (muscle), lateral head (→ 291)

10 Obturator externus (muscle)
11 Iliopsoas (muscle) (→ 206)
12 Gluteus medius (muscle) (→ 281)
13 Pectineus (muscle) (→ 267)
14 Quadratus femoris (muscle) (→ 282)
15 Adductor brevis (muscle) (→ 271)
16 Adductor magnus (muscle) (→ 281)
17 Gluteus maximus (muscle) (→ 268)
18 Vastus lateralis (muscle) (→ 282)
19 Vastus intermedius (muscle) (→ 268)
20 Vastus medialis (muscle) (→ 268)
21 Adductor longus (muscle) (→ 267)
22 Biceps femoris (muscle) (→ 281)
23 Adductor magnus (muscle) (→ 281)
24 Gastrocnemius (muscle), medial head (→ 291)
25 Plantaris (muscle) (→ 291)
26 Gastrocnemius (muscle), lateral head (→ 291)
27 Popliteus (muscle) (→ 292)

12.2.9 Lymph Nodes of Inguinal Area

1 Superficial inguinal nodes

They are located along the inguinal ligament and the greater saphenous vein (→ 263). They receive lymph from:

- Trunk wall below the navel
- Skin of outer genitals, perineum, buttocks, anus and parts of vagina
- Lateral angle of uterus, fundus of uterus
- Skin of leg
- Dense network of lymphatic vessels located in the leg, which run approximately with the greater saphenous vein. Drainage occurs via No. 2.

2 Deep inguinal nodes

(Rosenmüller's node)
Receives inflow originating from deep lymphatic vessels and from No. 1.

Drainage occurs via the **external iliac nodes** (→ 196).

In addition, there are the **superficial popliteal nodes**, located in the hollow of the knee (poplitea) near the small saphenous vein, which receive lymph from the lateral border of the foot and from the heel. Further there are the **deep popliteal nodes**, which receive inflow from the backside of the lower thigh. The variable anterior and posterior tibial nodes and the peritoneal nodes should also be mentioned here.

12.2.10 Superficial Vessels

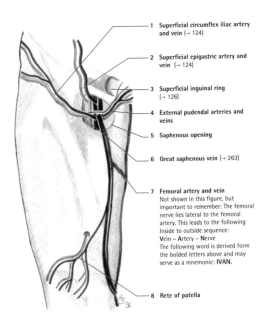

1 Superficial circumflex iliac artery and vein (→ 124)

2 Superficial epigastric artery and vein (→ 124)

3 Superficial inguinal ring (→ 126)

4 External pudendal arteries and veins

5 Saphenous opening

6 Great saphenous vein (→ 263)

7 Femoral artery and vein
Not shown in this figure, but important to remember: The femoral nerve lies lateral to the femoral artery. This leads to the following Inside to outside sequence:
Vein – Artery – Nerve
The following word is derived form the bolded letters above and may serve as a mnemonic: **IVAN.**

8 Rete of patella

12.2.11 Ligaments and Tendons of Knee Joint

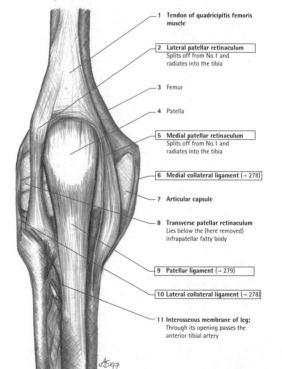

1 **Tendon of quadricipitis femoris muscle**

2 **Lateral patellar retinaculum**
Splits off from No.1 and radiates into the tibia

3 **Femur**

4 **Patella**

5 **Medial patellar retinaculum**
Splits off from No.1 and radiates into the tibia

6 **Medial collateral ligament (→ 278)**

7 **Articular capsule**

8 **Transverse patellar retinaculum**
Lies below the (here removed) infrapatellar fatty body

9 **Patellar ligament (→ 279)**

10 **Lateral collateral ligament (→ 278)**

11 **Interosseous membrane of leg;**
Through its opening passes the anterior tibial artery

12.2.12 Ligaments of Knee Joint

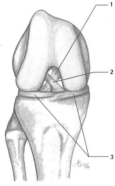

1 **Posterior cruciate ligament**
It passes **from mediosuperior to lateroinferior,** between the **intercondylar area** of tibia and the **intercondylar fossa** of femur. Together with the anterior cruciate ligament it prevents the displacement of the femur on the tibia and vice versa. It also acts as the main stabilizer during locomotion.

2 **Anterior cruciate ligament**
It too runs between the intercondylar area of tibia and the intercondylar fossa of femur, but prior to the posterior cruciate ligament and **from laterosuperior to medioinferior.**
Function: see above

The cruciate ligaments wrap around each other during medial rotation, and separate from each other during lateral rotation.

3 **Medial and lateral meniscus** (→ 279)

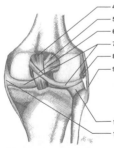

4 **Posterior cruciate ligament**

5 **Anterior cruciate ligament**

6 **Posterior meniscofemoral ligament**

7 **Medialis and lateral meniscus**

8 **Tendon of popliteus muscle** (→ 292)

9 **Fibular collateral ligament** (→ 277)
The fibular collateral ligament passes from the lateral epicondyle of the femur to the head of the fibula as a well defined cord. The collateral ligaments suppress ab- and adduction and stabilize the knee joint.

10 **Posterior ligament of the head of the fibula**

11 **Tibial collateral ligament** (→ 277)
It extends from the medial epicondyle of femur to the medial end and medial shaft of the tibia. It is firmly attached to the joint capsule and the medial meniscus.

Clinical Information
Most ligament ruptures involve the tibial collateral ligament.

12.2.13 Knee Joint and Meniscus

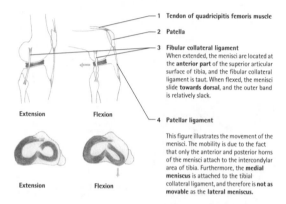

Extension   Flexion

Extension   Flexion

1 **Tendon of quadricipitis femoris muscle**

2 **Patella**

3 **Fibular collateral ligament**
When extended, the menisci are located at the **anterior part** of the superior articular surface of tibia, and the fibular collateral ligament is taut. When flexed, the menisci slide **towards dorsal**, and the outer band is relatively slack.

4 **Patellar ligament**

This figure illustrates the movement of the menisci. The mobility is due to the fact that only the anterior and posterior horns of the menisci attach to the intercondylar area of tibia. Furthermore, the **medial meniscus** is attached to the tibial collateral ligament, and therefore is **not as movable** as the **lateral meniscus**.

The knee joint, **articulatio genus**, is a combination of both hinge and pivot joint (rotational movements are possible while knees are bent!). The fibula is not part of this joint.

The menisci are C-shaped structures consisting of **fibrous cartilage**. Distinguished are the semicircle-shaped **medial meniscus** and the nearly circular **lateral meniscus**. Both, the lateral and medial meniscus, insert at one side at the anterior intercondylar area, at the other side at the posterior intercondylar area of tibia. They move together during knee movements (see above).
Their function is to ensure equal **pressure distribution** over the superior articular surface of tibia (the articular surface of the tibia is larger than that of the femoral condyles!). In addition they subdivide the joint cavity incompletely.

Clinical Information

Clinically very important is the fact, that the medial meniscus is connected via the joint capsule to the tibial collateral ligament. This structure is more frequently affected by injury, since ligament ruptures mainly involve the tibial collateral ligament.

12.2.14 Cutaneous Nerves of Anterior Femoral Region

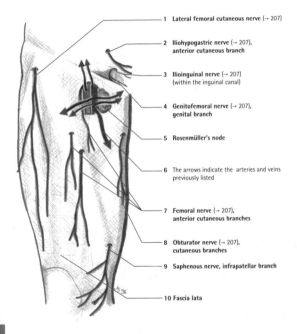

1 Lateral femoral cutaneous nerve (→ 207)

2 Iliohypogastric nerve (→ 207), anterior cutaneous branch

3 Ilioinguinal nerve (→ 207) (within the inguinal canal)

4 Genitofemoral nerve (→ 207), genital branch

5 Rosenmüller's node

6 The arrows indicate the arteries and veins previously listed

7 Femoral nerve (→ 207), anterior cutaneous branches

8 Obturator nerve (→ 207), cutaneous branches

9 Saphenous nerve, infrapatellar branch

10 Fascia lata

12.3 Posterior Femoral Region
12.3.1 Dorsal View of Muscles of Thigh I

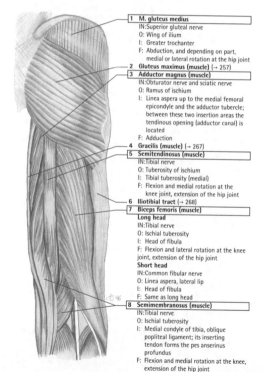

1 M. gluteus medius
IN: Superior gluteal nerve
O: Wing of ilium
I: Greater trochanter
F: Abduction, and depending on part, medial or lateral rotation at the hip joint

2 Gluteus maximus (muscle) (→ 257)

3 Adductor magnus (muscle)
IN: Obturator nerve and sciatic nerve
O: Ramus of ischium
I: Linea aspera up to the medial femoral epicondyle and the adductor tubercle; between these two insertion areas the tendinous opening (adductor canal) is located
F: Adduction

4 Gracilis (muscle) (→ 267)

5 Semitendinosus (muscle)
IN: Tibial nerve
O: Tuberosity of ischium
I: Tibial tuberosity (medial)
F: Flexion and medial rotation at the knee joint, extension of the hip joint

6 Iliotibial tract (→ 268)

7 Biceps femoris (muscle)
Long head
IN: Tibial nerve
O: Ischial tuberosity
I: Head of fibula
F: Flexion and lateral rotation at the knee joint, extension of the hip joint
Short head
IN: Common fibular nerve
O: Linea aspera, lateral lip
I: Head of fibula
F: Same as long head

8 Semimembranosus (muscle)
IN: Tibial nerve
O: Ischial tuberosity
I: Medial condyle of tibia, oblique popliteal ligament; its inserting tendon forms the pes anserinus profundus
F: Flexion and medial rotation at the knee, extension of the hip joint

12.3.2 Dorsal View of Muscles of Thigh II

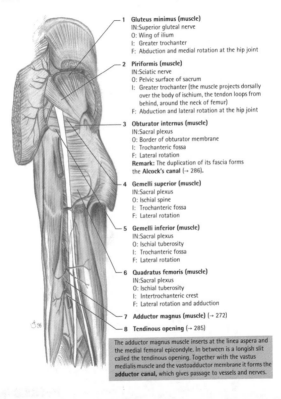

1 **Gluteus minimus (muscle)**
IN: Superior gluteal nerve
O: Wing of ilium
I: Greater trochanter
F: Abduction and medial rotation at the hip joint

2 **Piriformis (muscle)**
IN: Sciatic nerve
O: Pelvic surface of sacrum
I: Greater trochanter (the muscle projects dorsally over the body of ischium, the tendon loops from behind, around the neck of femur)
F: Abduction and lateral rotation at the hip joint

3 **Obturator internus (muscle)**
IN: Sacral plexus
O: Border of obturator membrane
I: Trochanteric fossa
F: Lateral rotation
Remark: The duplication of its fascia forms the **Alcock's canal** (→ 286).

4 **Gemelli superior (muscle)**
IN: Sacral plexus
O: Ischial spine
I: Trochanteric fossa
F: Lateral rotation

5 **Gemelli inferior (muscle)**
IN: Sacral plexus
O: Ischial tuberosity
I: Trochanteric fossa
F: Lateral rotation

6 **Quadratus femoris (muscle)**
IN: Sacral plexus
O: Ischial tuberosity
I: Intertrochanteric crest
F: Lateral rotation and adduction

7 **Adductor magnus (muscle)** (→ 272)

8 **Tendinous opening** (→ 285)

The adductor magnus muscle inserts at the linea aspera and the medial femoral epicondyle. In between is a longish slit called the tendinous opening. Together with the vastus medialis muscle and the vastoadductor membrane it forms the **adductor canal**, which gives passage to vessels and nerves.

12.3.3 Posterior Femoral Region

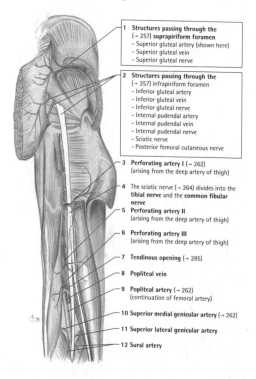

1 Structures passing through the (→ 257) **suprapiriform foramen**
- Superior gluteal artery (shown here)
- Superior gluteal vein
- Superior gluteal nerve

2 Structures passing through the (→ 257) **infrapiriform foramen**
- Inferior gluteal artery
- Inferior gluteal vein
- Inferior gluteal nerve
- Internal pudendal artery
- Internal pudendal vein
- Internal pudendal nerve
- Sciatic nerve
- Posterior femoral cutaneous nerve

3 Perforating artery I (→ 262)
(arising from the deep artery of thigh)

4 The sciatic nerve (→ 264) divides into the **tibial nerve** and the **common fibular nerve**

5 Perforating artery II
(arising from the deep artery of thigh)

6 Perforating artery III
(arising from the deep artery of thigh)

7 Tendinous opening (→ 285)

8 Popliteal vein

9 Popliteal artery (→ 262)
(continuation of femoral artery)

10 Superior medial genicular artery (→ 262)

11 Superior lateral genicular artery

12 Sural artery

12.3.4 Muscles of Posterior Femoral Region

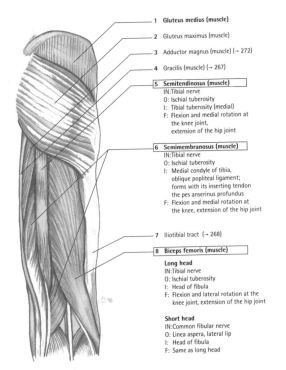

1 **Gluteus medius (muscle)**

2 Gluteus maximus (muscle)

3 Adductor magnus (muscle) (→ 272)

4 Gracilis (muscle) (→ 267)

5 **Semitendinosus (muscle)**
IN: Tibial nerve
O: Ischial tuberosity
I: Tibial tuberosity (medial)
F: Flexion and medial rotation at
   the knee joint,
   extension of the hip joint

6 **Semimembranosus (muscle)**
IN: Tibial nerve
O: Ischial tuberosity
I: Medial condyle of tibia,
   oblique popliteal ligament;
   forms with its inserting tendon
   the pes anserinus profundus
F: Flexion and medial rotation at
   the knee, extension of the hip joint

7 Iliotibial tract (→ 268)

8 **Biceps femoris (muscle)**

Long head
IN: Tibial nerve
O: Ischial tuberosity
I: Head of fibula
F: Flexion and lateral rotation at the
   knee joint, extension of the hip joint

Short head
IN: Common fibular nerve
O: Linea aspera, lateral lip
I: Head of fibula
F: Same as long head

12.3.5 Topography of Posterior Femoral Region

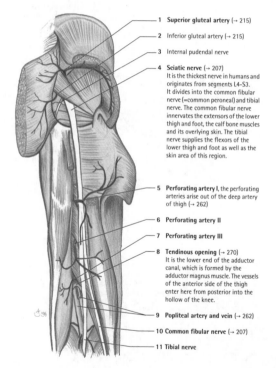

1 **Superior gluteal artery** (→ 215)

2 **Inferior gluteal artery** (→ 215)

3 **Internal pudendal nerve**

4 **Sciatic nerve** (→ 207)
It is the thickest nerve in humans and originates from segments L4–S3. It divides into the common fibular nerve (=common peroneal) and tibial nerve. The common fibular nerve innervates the extensors of the lower thigh and foot, the calf bone muscles and its overlying skin. The tibial nerve supplies the flexors of the lower thigh and foot as well as the skin area of this region.

5 **Perforating artery I**, the perforating arteries arise out of the deep artery of thigh (→ 262)

6 **Perforating artery II**

7 **Perforating artery III**

8 **Tendinous opening** (→ 270)
It is the lower end of the adductor canal, which is formed by the adductor magnus muscle. The vessels of the anterior side of the thigh enter here from posterior into the hollow of the knee.

9 **Popliteal artery and vein** (→ 262)

10 **Common fibular nerve** (→ 207)

11 **Tibial nerve**

12.3.6 Alcock's Canal

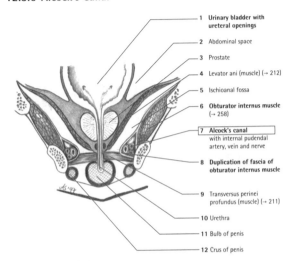

1. Urinary bladder with ureteral openings
2. Abdominal space
3. Prostate
4. Levator ani (muscle) (→ 212)
5. Ischioanal fossa
6. Obturator internus muscle (→ 258)
7. **Alcock's canal** with internal pudendal artery, vein and nerve
8. **Duplication of fascia of obturator internus muscle**
9. Transversus perinei profundus (muscle) (→ 211)
10. Urethra
11. Bulb of penis
12. Crus of penis

The **pudendal canal (Alcock's canal)** lies within the ischioanal fossa and is formed by the duplication of the fascia of the obturator internus muscle. It gives passage along the ischial tuberosity to the internal pudendal vessels and the pudendal nerve.

It should be noted, that the internal pudendal artery, vein and nerve leave the pelvis through the infrapiriform foramen (→ 257), embrace the sacrospinal ligament, and then enter the Alcock's canal by returning back into the pelvic space through the lesser sciatic foramen. The ischioanal fossa contains the fat body of ischioanal fossa.

The **pudendal nerve**, the most downstream branch of the sacral plexus, supplies sensory innervation to the skin of anus, perineum and genitals, motor innervation to the muscles of the pelvic floor, sphincter muscle of urethra and external sphincter muscle of anus, and vegetative innervation to the genitals.

12.3.7 Popliteal Fossa

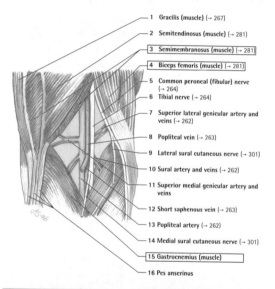

1 Gracilis (muscle) (→ 267)

2 Semitendinosus (muscle) (→ 281)

3 Semimembranosus (muscle) (→ 281)

4 Biceps femoris (muscle) (→ 281)

5 Common peroneal (fibular) nerve (→ 264)

6 Tibial nerve (→ 264)

7 Superior lateral genicular artery and veins (→ 262)

8 Popliteal vein (→ 263)

9 Lateral sural cutaneous nerve (→ 301)

10 Sural artery and veins (→ 262)

11 Superior medial genicular artery and veins

12 Short saphenous vein (→ 263)

13 Popliteal artery (→ 262)

14 Medial sural cutaneous nerve (→ 301)

15 Gastrocnemius (muscle)

16 Pes anserinus

The **popliteal fossa** is a rhomboid area bordered by the biceps femoris, semimembranosus and gastrocnemius muscles. It contains a fatty body, which appears as protrusion during extension of the joint. The hollow of the knee is traversed by the following tracts, connecting upper and lower thigh: The common fibular and tibial nerve, the popliteal artery and vein, the lymphatic vessels and smaller nerves and blood vessels.

The tendons of the sartorius, gracilis and semitendinosus muscles together form the **pes anserinus,** which inserts proximally and medially into the tibia. Since this structure bears resemblance to the foot of a goose, it received this name.

12.4 Lower Thigh
12.4.1 Osteofibrous Tunnels of Lower Thigh

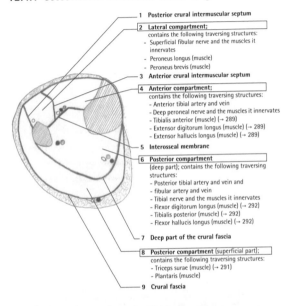

1 **Posterior crural intermuscular septum**

2 **Lateral compartment;**
   contains the following traversing structures:
   - Superficial fibular nerve and the muscles it innervates
   - Peroneus longus (muscle)
   - Peroneus brevis (muscle)

3 **Anterior crural intermuscular septum**

4 **Anterior compartment;**
   contains the following traversing structures:
   - Anterior tibial artery and vein
   - Deep peroneal nerve and the muscles it innervates
   - Tibialis anterior (muscle) (→ 289)
   - Extensor digitorum longus (muscle) (→ 289)
   - Extensor hallucis longus (muscle) (→ 289)

5 **Interosseal membrane**

6 **Posterior compartment**
   (deep part); contains the following traversing structures:
   - Posterior tibial artery and vein and
   - fibular artery and vein
   - Tibial nerve and the muscles it innervates
   - Flexor digitorum longus (muscle) (→ 292)
   - Tibialis posterior (muscle) (→ 292)
   - Flexor hallucis longus (muscle) (→ 292)

7 **Deep part of the crural fascia**

8 **Posterior compartment** (superficial part);
   contains the following traversing structures:
   - Triceps surae (muscle) (→ 291)
   - Plantaris (muscle)

9 **Crural fascia**

Clinical Information
Compartment syndrome is characterized by increased tissue pressure within the enclosed muscular compartment. The resulting reduction in blood flow leads to neuromuscular failure and muscle necrosis. This occurrence may be e.g. trauma-induced (bleeding) or the result of inflammation (edema).

The **therapy** consists of cutting away the fascia (fasciotomy) to relieve tension or pressure.

12.4.2 Ventral View of Lower Thigh Muscles

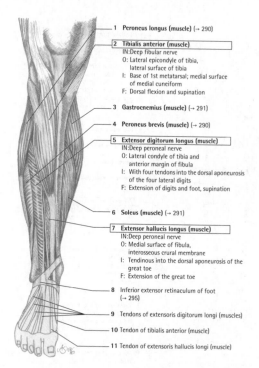

1 **Peroneus longus (muscle)** (→ 290)

2 **Tibialis anterior (muscle)**
IN: Deep fibular nerve
O: Lateral epicondyle of tibia,
lateral surface of tibia
I: Base of 1st metatarsal; medial surface
of medial cuneiform
F: Dorsal flexion and supination

3 **Gastrocnemius (muscle)** (→ 291)

4 **Peroneus brevis (muscle)** (→ 290)

5 **Extensor digitorum longus (muscle)**
IN: Deep peroneal nerve
O: Lateral condyle of tibia and
anterior margin of fibula
I: With four tendons into the dorsal aponeurosis
of the four lateral digits
F: Extension of digits and foot, supination

6 **Soleus (muscle)** (→ 291)

7 **Extensor hallucis longus (muscle)**
IN: Deep peroneal nerve
O: Medial surface of fibula,
interosseous crural membrane
I: Tendinous into the dorsal aponeurosis of the
great toe
F: Extension of the great toe

8 **Inferior extensor retinaculum of foot**
(→ 295)

9 **Tendons of extensoris digitorum longi (muscles)**

10 **Tendon of tibialis anterior (muscle)**

11 **Tendon of extensoris hallucis longi (muscle)**

12.4.3 Lateral View of Lower Thigh Muscles

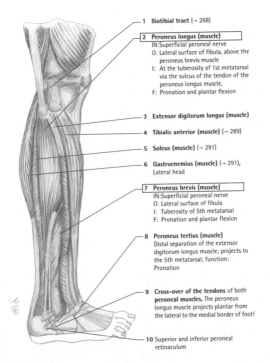

1 **Iliotibial tract** (→ 268)

2 **Peroneus longus (muscle)**
IN: Superficial peroneal nerve
O: Lateral surface of fibula, above the
   peroneus brevis muscle
I: At the tuberosity of 1st metatarsal
   via the sulcus of the tendon of the
   peroneus longus muscle,
F: Pronation and plantar flexion

3 **Extensor digitorum longus (muscle)**

4 **Tibialis anterior (muscle)** (→ 289)

5 **Soleus (muscle)** (→ 291)

6 **Gastrocnemius (muscle)** (→ 291),
   Lateral head

7 **Peroneus brevis (muscle)**
IN: Superficial peroneal nerve
O: Lateral surface of fibula
I: Tuberosity of 5th metatarsal
F: Pronation and plantar flexion

8 **Peroneus tertius (muscle)**
Distal separation of the extensor
digitorum longus muscle; projects to
the 5th metatarsal; function:
Pronation

9 **Cross-over of the tendons** of both
peroneal muscles. The peroneus
longus muscle projects plantar from
the lateral to the medial border of foot!

10 Superior and inferior peroneal
retinaculum

12.4.4 Dorsal View of Lower Thigh Muscles I

1 **Gastrocnemius (muscle), medial head**
IN: Tibial nerve
O: Medial epicondyle of femur
I: At the tuberosity of calcaneus via the calcaneal tendon
F: Plantar flexion at the upper and supination at the lower ankle joint, flexion at the knee joint

2 **Gastrocnemius (muscle), lateral head**
IN: Tibial nerve
O: Lateral epicondyle of femur
I: At the tuberosity of calcaneus via the calcaneal tendon
F: Plantar flexion at the upper and supination at the lower ankle joint, flexion at the knee joint

3 Arcuate popliteal ligament

4 Oblique popliteal ligament

5 Popliteal artery and vein (→ 287)
within the tendinous arch of soleus muscle

6 Plantaris (muscle) (variation)

7 **Soleus (muscle)**
IN: Tibial nerve
O: Posterior aspect and head of fibula, soleal line; the tendinous arch of the soleus muscle is located in-between
I: At the tuberosity of calcaneus via the calcaneal tendon
F: Plantar flexion at the upper and supination at the lower ankle joint, flexion at the knee joint

8 Calcaneal tendon, Achilles tendon (→ 294)

9 Flexor retinaculum of foot

10 Flexor digitorum longus (muscle) (→ 292)

The deep part of crural fascia (→ 288) separates the superficial layer of lower thigh flexors from the deep layer (osteofibrous tunnels).

Clinical Information

1, 2 and 7 together form the triceps surae muscle. Paralysis of the tibial nerve leads to the development of a pes calcaneus.

12.4.5 Dorsal View of Lower Thigh Muscles II

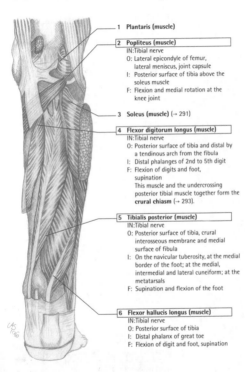

1 **Plantaris (muscle)**

2 **Popliteus (muscle)**
IN: Tibial nerve
O: Lateral epicondyle of femur,
   lateral meniscus, joint capsule
I: Posterior surface of tibia above the
   soleus muscle
F: Flexion and medial rotation at the
   knee joint

3 **Soleus (muscle) (→ 291)**

4 **Flexor digitorum longus (muscle)**
IN: Tibial nerve
O: Posterior surface of tibia and distal by
   a tendinous arch from the fibula
I: Distal phalanges of 2nd to 5th digit
F: Flexion of digits and foot,
   supination
   This muscle and the undercrossing
   posterior tibial muscle together form the
   crural chiasm (→ 293).

5 **Tibialis posterior (muscle)**
IN: Tibial nerve
O: Posterior surface of tibia, crural
   interosseous membrane and medial
   surface of fibula
I: On the navicular tuberosity, at the medial
   border of the foot; at the medial,
   intermedial and lateral cuneiform; at the
   metatarsals
F: Supination and flexion of the foot

6 **Flexor hallucis longus (muscle)**
IN: Tibial nerve
O: Posterior surface of tibia
I: Distal phalanx of great toe
F: Flexion of digit and foot, supination

12.4.6 Muscle Insertions and Origins of Lower Thigh

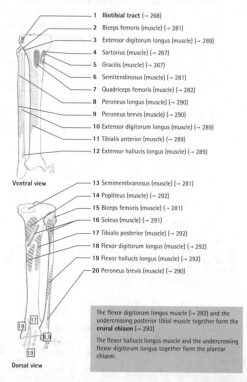

1 Iliotibial tract (→ 268)
2 Biceps femoris (muscle) (→ 281)
3 Extensor digitorum longus (muscle) (→ 289)
4 Sartorius (muscle) (→ 267)
5 Gracilis (muscle) (→ 267)
6 Semitendinosus (muscle) (→ 281)
7 Quadriceps femoris (muscle) (→ 282)
8 Peroneus longus (muscle) (→ 290)
9 Peroneus brevis (muscle) (→ 290)
10 Extensor digitorum longus (muscle) (→ 289)
11 Tibialis anterior (muscle) (→ 289)
12 Extensor hallucis longus (muscle) (→ 289)

Ventral view

13 Semimembranosus (muscle) (→ 281)
14 Popliteus (muscle) (→ 292)
15 Biceps femoris (muscle) (→ 281)
16 Soleus (muscle) (→ 291)
17 Tibialis posterior (muscle) (→ 292)
18 Flexor digitorum longus (muscle) (→ 292)
19 Flexor hallucis longus (muscle) (→ 292)
20 Peroneus brevis (muscle) (→ 290)

Dorsal view

The flexor digitorum longus muscle (→ 292) and the undercrossing posterior tibial muscle together form the **crural chiasm** (→ 292)

The flexor hallucis longus muscle and the undercrossing flexor digitorum longus together form the plantar chiasm.

12.5   Foot, Ankle Joint

12.5.1 Ligament System of Foot and Ankle Joint

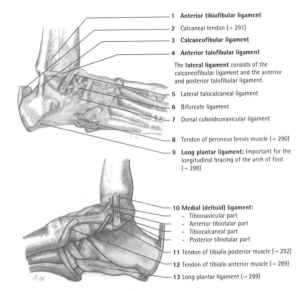

1 Anterior tibiofibular ligament

2 Calcaneal tendon (→ 291)

3 Calcaneofibular ligament

4 Anterior talofibular ligament

The **lateral ligament** consists of the calcaneofibular ligament and the anterior and posterior talofibular ligament.

5 Lateral talocalcaneal ligament

6 Bifurcate ligament

7 Dorsal cuboideonavicular ligament

8 Tendon of peroneus brevis muscle (→ 290)

9 **Long plantar ligament;** Important for the longitudinal bracing of the arch of foot (→ 299)

10 Medial (deltoid) ligament:
  - Tibionavicular part
  - Anterior tibiotalar part
  - Tibiocalcaneal part
  - Posterior tibiotalar part

11 Tendon of tibialis posterior muscle (→ 292)

12 Tendon of tibialis anterior muscle (→ 289)

13 Long plantar ligament (→ 299)

As the medial ligament, the lateral ligament stabilizes the **upper ankle joint**. Both have a fanlike structure, however the lateral ligament is not as strong as the medial one. Through their action at the upper ankle joint (articulatio talocruralis), both ligaments prevent canting of the foot. In any position (the articulatio talocruralis permits flexion and extension) portions of the ligaments are under tension.
Remember: The **lower ankle joint** permits the following movements: Supination and pronation of the foot.

12.5.2 Musculature of Back of Foot I

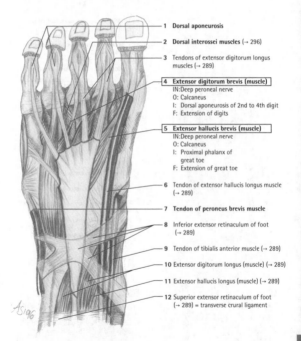

1 **Dorsal aponeurosis**

2 **Dorsal interossei muscles** (→ 296)

3 Tendons of extensor digitorum longus muscles (→ 289)

4 **Extensor digitorum brevis (muscle)**
   IN: Deep peroneal nerve
   O: Calcaneus
   I: Dorsal aponeurosis of 2nd to 4th digit
   F: Extension of digits

5 **Extensor hallucis brevis (muscle)**
   IN: Deep peroneal nerve
   O: Calcaneus
   I: Proximal phalanx of great toe
   F: Extension of great toe

6 Tendon of extensor hallucis longus muscle (→ 289)

7 **Tendon of peroneus brevis muscle**

8 Inferior extensor retinaculum of foot (→ 289)

9 Tendon of tibialis anterior muscle (→ 289)

10 Extensor digitorum longus (muscle) (→ 289)

11 Extensor hallucis longus (muscle) (→ 289)

12 Superior extensor retinaculum of foot (→ 289) = transverse crural ligament

12.5.3 Musculature of Back of Foot II

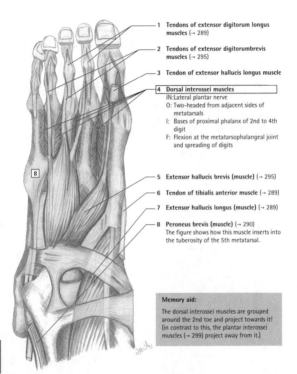

1 **Tendons of extensor digitorum longus muscles** (→ 289)

2 **Tendons of extensor digitorumbrevis muscles** (→ 295)

3 **Tendon of extensor hallucis longus muscle**

4 **Dorsal interossei muscles**
   IN: Lateral plantar nerve
   O: Two-headed from adjacent sides of metatarsals
   I: Bases of proximal phalanx of 2nd to 4th digit
   F: Flexion at the metatarsophalangeal joint and spreading of digits

5 **Extensor hallucis brevis (muscle)** (→ 295)

6 **Tendon of tibialis anterior muscle** (→ 289)

7 **Extensor hallucis longus (muscle)** (→ 289)

8 **Peroneus brevis (muscle)** (→ 290)
   The figure shows how this muscle inserts into the tuberosity of the 5th metatarsal.

Memory aid:

The dorsal interossei muscles are grouped around the 2nd toe and project towards it! (in contrast to this, the plantar interossei muscles (→ 299) project away from it.)

12.5.4 Plantar View of Muscles of Foot (Superficial Layer)

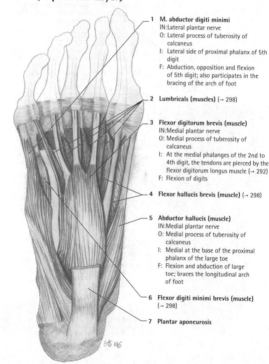

1 **M. abductor digiti minimi**
IN: Lateral plantar nerve
O: Lateral process of tuberosity of calcaneus
I: Lateral side of proximal phalanx of 5th digit
F: Abduction, opposition and flexion of 5th digit; also participates in the bracing of the arch of foot

2 **Lumbricals (muscles)** (→ 298)

3 **Flexor digitorum brevis (muscle)**
IN: Medial plantar nerve
O: Medial process of tuberosity of calcaneus
I: At the medial phalanges of the 2nd to 4th digit, the tendons are pierced by the flexor digitorum longus muscle (→ 292)
F: Flexion of digits

4 **Flexor hallucis brevis (muscle)** (→ 298)

5 **Abductor hallucis (muscle)**
IN: Medial plantar nerve
O: Medial process of tuberosity of calcaneus
I: Medial at the base of the proximal phalanx of the large toe
F: Flexion and abduction of large toe; braces the longitudinal arch of foot

6 **Flexor digiti minimi brevis (muscle)** (→ 298)

7 **Plantar aponeurosis**

12.5.5 Plantar View of Muscles of Foot (Medial Layer)

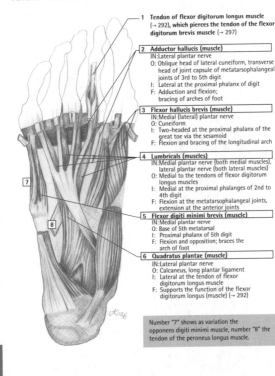

1 **Tendon of flexor digitorum longus muscle**
   (→ 292), which pierces the tendon of the flexor
   digitorum brevis muscle (→ 297)

2 **Adductor hallucis (muscle)**
   IN: Lateral plantar nerve
   O: Oblique head of lateral cuneiform, transverse
      head of joint capsule of metatarsophalangeal
      joints of 3rd to 5th digit
   I: Lateral at the proximal phalanx of digit
   F: Adduction and flexion;
      bracing of arches of foot

3 **Flexor hallucis brevis (muscle)**
   IN: Medial (lateral) plantar nerve
   O: Cuneiform
   I: Two-headed at the proximal phalanx of the
      great toe via the sesamoid
   F: Flexion and bracing of the longitudinal arch

4 **Lumbricals (muscles)**
   IN: Medial plantar nerve (both medial muscles),
       lateral plantar nerve (both lateral muscles)
   O: Medial to the tendons of flexor digitorum
      longus muscles
   I: Medial at the proximal phalanges of 2nd to
      4th digit
   F: Flexion at the metatarsophalangeal joints,
      extension at the anterior joints

5 **Flexor digiti minimi brevis (muscle)**
   IN: Medial plantar nerve
   O: Base of 5th metatarsal
   I: Proximal phalanx of 5th digit
   F: Flexion and opposition; braces the
      arch of foot

6 **Quadratus plantae (muscle)**
   IN: Lateral plantar nerve
   O: Calcaneus, long plantar ligament
   I: Lateral at the tendon of flexor digitorum
      longus muscle
   F: Supports the function of the flexor
      digitorum longus (muscle) (→ 292)

Number "7" shows as variation the
opponens digiti minimi muscle, number "8" the
tendon of the peroneus longus muscle.

12.5.6 Plantar View of Muscles of Foot (Deep Layer)

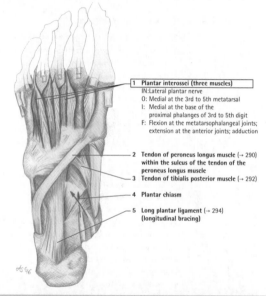

1 **Plantar interossei (three muscles)**
IN: Lateral plantar nerve
O: Medial at the 3rd to 5th metatarsal
I: Medial at the base of the
  proximal phalanges of 3rd to 5th digit
F: Flexion at the metatarsophalangeal joints;
  extension at the anterior joints; adduction

2 **Tendon of peroneus longus muscle (→ 290)
within the sulcus of the tendon of the
peroneus longus muscle**

3 **Tendon of tibialis posterior muscle (→ 292)**

4 **Plantar chiasm**

5 **Long plantar ligament (→ 294)
(longitudinal bracing)**

The **plantar interossei muscles I–III** are oriented away from the 2nd toe! In this figure they are shown darker than the dorsal interossei muscles.

Arch of foot
The skeleton of foot is constructed so that it forms an angle, with its one side formed by the calcaneus and corpus tali and the other side formed by the toes and the cuneiform, navicular, cuboid and metatarsal bones. This anterior part makes up a concave groove, the **longitudinal** and **transverse arch**. Both are maintained through the action of ligaments and muscles. If parts of this construction

fail, orthopedic abnormalities such as flat and splay foot can be the result.

Pronation and supination
Pronation of the foot means to depress the medial border of the foot together with abduction and dorsal flexion. In contrast to this, supination means to elevate the medial border of the foot with concurrent adduction and plantar flexion.

The **plantar chiasm** is formed by the tendons of the flexor digitorum longus and the undercrossing flexor hallucis longus muscle.

12.5.7 Muscle Insertions and Origins of Foot

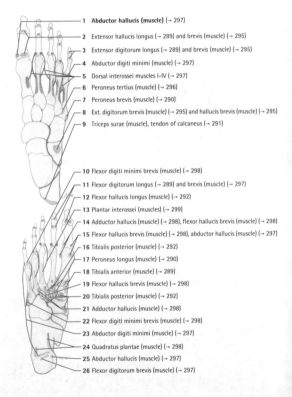

1 Abductor hallucis (muscle) (→ 297)
2 Extensor hallucis longus (→ 289) and brevis (muscle) (→ 295)
3 Extensor digitorum longus (→ 289) and brevis (muscle) (→ 295)
4 Abductor digiti minimi (muscle) (→ 297)
5 Dorsal interossei muscles I–IV (→ 297)
6 Peroneus tertius (muscle) (→ 296)
7 Peroneus brevis (muscle) (→ 290)
8 Ext. digitorum brevis (muscle) (→ 295) and hallucis brevis (muscle) (→ 295)
9 Triceps surae (muscle), tendon of calcaneus (→ 291)

10 Flexor digiti minimi brevis (muscle) (→ 298)
11 Flexor digitorum longus (→ 289) and brevis (muscle) (→ 297)
12 Flexor hallucis longus (muscle) (→ 292)
13 Plantar interossei (muscles) (→ 299)
14 Adductor hallucis (→ 298), flexor hallucis brevis (muscle) (→ 298)
15 Flexor hallucis brevis (muscle) (→ 298), abductor hallucis (muscle) (→ 297)
16 Tibialis posterior (muscle) (→ 292)
17 Peroneus longus (muscle) (→ 290)
18 Tibialis anterior (muscle) (→ 289)
19 Flexor hallucis brevis (muscle) (→ 298)
20 Tibialis posterior (muscle) (→ 292)
21 Adductor hallucis (muscle) (→ 298)
22 Flexor digiti minimi brevis (muscle) (→ 298)
23 Abductor digiti minimi (muscle) (→ 297)
24 Quadratus plantae (muscle) (→ 298)
25 Abductor hallucis (muscle) (→ 297)
26 Flexor digitorum brevis (muscle) (→ 297)

12.6 Cutaneous Nerves, Dermatomes
12.6.1 Cutaneous Nerves of Lower Limb, Dermatomes

Cutaneous nerves of leg from posterior (→ 252):
From superior to inferior the **gluteal region** is supplied by the
- Superior cluneal nerves (lumbar nerves)
- Medial cluneal nerves (sacral nerves)
- Inferior cluneal nerves (arising from the posterior femoral cutaneous nerve).

The **area below the iliac crest** is innervated by the iliohypogastric nerve.
Supply of the **posterior femoral region** is provided
- for the medial border by the obturator nerve
- for the lateral border by the lateral femoral cutaneous nerve
- for all other areas by the posterior femoral cutaneous nerve.

The **backside of the lower thigh** is supplied medial to the saphenous nerve and lateral to the superficial peroneal nerve by the sural cutaneous nerve. The **area of the Achilles tendon and above** is supplied by the sural nerve.

Cutaneous nerves of leg from anterior:
The **front side of the thigh** is supplied laterally, by the lateral femoral cutaneous nerve and medially by the femoral nerve. The **adductor area** is supplied by the obturator nerve. The region surrounding the saphenous opening as well as the urogenital region (femoral and genital region) are under the control of the genitofemoral nerve. The **pubic region** is supplied by the ilioinguinal nerve.

Supply of the **front side of the lower thigh** is provided by two nerves: The medial part is innervated by the saphenous nerve, the lateral part by the common peroneal nerve.

The **back of foot** is the area, lateral to the zone that is supplied by the sural nerve, which is innervated by the superficial peroneal nerve.

The **sole of foot** still belongs to the supply area of the tibial nerve.
The dermatomes of the leg are supplied by segments L1 (area of the iliac crest) through S5 (gluteal cleft).
The course of the borders between dermatomes are the result of sprouting and growth processes during embryonic development.

Medical Spanish pocket plus

Medical Spanish pocket plus
ISBN 1-59103-213-X, US $ 22.95

- Includes contents of both
  Medical Spanish pocket and
  **Medical Spanish Dictionary
  pocket**

- Vital communication tool

- Clearly organized by situation:
  interview, examination, course
  of visit

- Accurate translations for almost
  every health-related term

- Bilingual dictionary contains
  medical terminology specific to
  Mexico, Puerto Rico, Cuba and
  other countries

Also available as PDA software!

**Medical Spanish pocket plus
for PDA** (2487 kb)

**Medical Spanish Dictionary
pocket for PDA** (2343 kb)

**Medical Spanish pocket
for PDA** (670 kb)

 Table of Contents, Text samples, PDA demo files... www.media4u .com

Drug pocket plus

Drug pocket plus
ISBN 1-59103-226-1; US $ 24.95

- Includes contents of both **Drug pocket** and **Drug Therapy pocket**

- All major US drugs with brand names, forms, strengths and dosages

- Brief descriptions of mechanisms of action, effects, side effects and contraindications

- Details on elimination half-lives, pregnancy risk categories and use during lactation

- A comprehensive system of cross-references to create bridges between the information on drug therapy for a given disease and the information on a particular drug

- Completely indexed for quick access

Drug pocket for PDA (460 kb)
Drug Therapy pocket for PDA (840 kb)

 Table of Contents, Text samples, PDA demo files... www.**media4u**.com

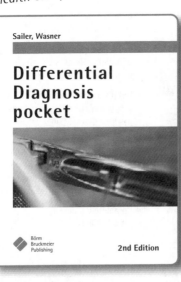

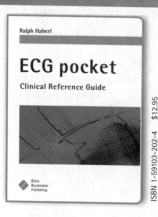

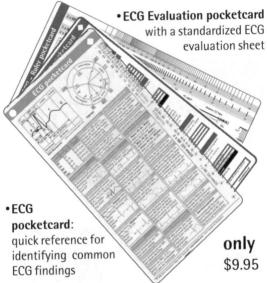

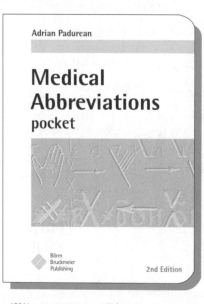

H&P pocketcard

History & Physical Exam pocketcard

- A brief survey of all main points of history and physical examination:

- ID, CC, HPI, PMH, allergies, current medications, SH, FH, sexual history

- GA, VS, MS, skin, lymph nodes, HEENT, neck, thorax, lungs, female breasts, CVS, abdomen, genitalia, rectum, musculo-skeletal, neurologic system

only
$3.95

- The History and Physical Exam pocketcard is a must-have companion for medical students beginning in medical practice

The Skeletal Muscles

Musculature of head

Masticatory musculature

Masseter muscle
- O zygomatic arch
- I masseteric tuberosity
- F elevates mandible, closes jaws
- IN masseteric nerve
- superficial part and strong deep part

Temporal muscle
- O temp. surface
- I coronoid proc. of mandible
- F strongest elevator of mandible
- IN deep temporal nerves

Lateral pterygoid muscle
- O lat. lamina of pterygoid proc., infratemporal surface of greater wing of sphenoid bone
- I pterygoid fovea; articular disc
- F involved in all movements of mouth opening; initiation of mouth opening
- IN lat. pterygoid nerve

Medial pterygoid muscle
- O pterygoid fossa
- I pterygoid tuberosity
- F mandible: elevate, protrude, lateral and rotatory movement
- IN med. pterygoid nerve

Infrahyoid musculature
- F elevation of the larynx, oral opening; fixation hyoid bone, head: FLX
- IN deep cervical ansa, thyrohyoid ra. (C1-C3)

Sternohyoid muscle
- O posterior surface of manubrium of sternum
- I lat. body of hyoid bone

Omohyoid muscle
- O inf. belly: sup. border of scapula
- sup. belly: lat. at the body of hyoid bone
- F keeps the internal jugular vein open

Sternothyroid muscle
- O posterior surface of manubrium of sternum
- I oblique line of thyroid cartilage

Thyrohyoid muscle
- O oblique line of thyroid cartilage
- I lat. at the body of hyoid bone

Autochthonous or genuine musculature of back
syn. erector spinae muscle
- F bilateral contraction: REC

Medial tract

Spinal system
- +F VC: lateral bending
- R u. caudal, a. cranial

Interspinales muscle
- O and I 2 adjacent spinal processes
- IN dorsal branches
- R (C1-Th3, Th11-L5)
- R especially at the CVC and LVC

Spinalis muscle
- O and I skip 1 or >1 spinal processes
- IN dorsal branches (C2-Th10)
- R especially at the TVC

Transversospinal system
- O transverse processes
- I spinous processes
- +F VC: lateral bending, ROT

Rotatores brevis muscles
- O and I to next higher spinal proc.
- IN dorsal branches (Th1-11)
- R especially at the TVC

Rotatores longi muscles
- O and I skip 1 segment
- IN dorsal branches (Th1-11)
- R especially at the TVC

Multifidi muscles
- O and I skip 2-4 segments
- IN dorsal branches (C3-S4)
- R especially at the LVC

Semispinalis muscle
- O and I skips >4 segments
- IN dorsal branches (C1-5, C3-6, Th4-6)
- R especially at the CVC

Lateral tract
- F VC: Lateral bending

Transversal system

Iliocostalis muscle :
- IN dorsal branches (C4-L3)

- Iliocostalis cervicis muscle
- O ribs 3-6
- I transverse processes CV 4-6

- Iliocostalis thoracis muscle
- O ribs 7-12
- I ribs 1-6

- Iliocostalis lumborum
- O sacrum, iliac crest
- I up to cst. processes ab. LV and ribs 4-7

Longissimus muscle:
- IN dorsal branches (C2-L5)

- Longissimus capitis muscle
- O trv. processes CV approx. 5-7, TV 1- approx. 5
- I mastoid proc.

- Longissimus cervicis muscle
- O trv. processes TV 1-5
- I trv. processes CV 2-5

- Longissimus thoracis muscle
- O sacrum, spn. procc. LV, trv. proc. lower TV
- I up to 1 or 2 rib

Intertransverse system
- O and I between trv. procc. CV 2-7
- IN dorsal branch (C1-6)

Spinotransversal system

Splenius capitis and cervicis muscle
- O spn. proc. 4 lower CV, 3 upper TV
- I capitis: Mastoid proc., cervicis: trv. proc. CV 1-2
- F head: ROT
- IN dorsal branches (C1-8)

Short deep muscles of back of neck
- F bilateral: head: REC
- IN suboccipital nerve (C1)

Rectus capitis posterior major muscle
- O spinous proc. of axis
- I occipital bone, inf. nuchal line
- +F head: ROT

Rectus capitis posterior minor muscle
- O post. tubercle of atlas
- I medial to the major muscle

Oblique capitis inferior muscle
- O spinous proc. of axis
- I trv. proc. of atlas
- +F head: ROT

Oblique capitis superior muscle
- O trv. proc. of atlas
- I occipital bone
- +F head: lateral bending

Pre-vertebral musculature

Rectus capitis lateralis muscle
- O trv. proc. of atlas
- I jugular proc. of occipital bone
- F head: lateral bending
- IN C1

Rectus capitis anterior muscle
- O lateral mass of atlas
- I basilar part of occipital bone
- F head: INC
- IN cervical pl. (C1)

Longus colli and capitis muscles
- O especially bo. of vertebrae, trv. procc. up to TV3
- I especially trv. procc. CVC, occipital bone
- F head: INC, lateral bending
- IN cervical and brachial pl. (C1-8)

Scalenus muscles
- O trv. proc. CVC
- F inspiration, CVC: lateral bending
- IN cervical and brachial pl. (C5-8)

- Scalenus anterior muscle
- I anterior at the 1st rib

- Scalenus medius muscle
- I further posterior at the 1st rib

- Scalenus posterior muscle
- I further posterior at the 2nd rib

Quadratus lumborum muscle
- O iliac crest
- I costal proc. LVC, 12th rib
- F VC: lateral bending
- IN Th12, L1-3

Respiratory musculature

Diaphragm
I tendinous center
F inspiration, abdominal compression
IN phrenic nerves (C3-5, mainly C4)
O sternal p.: (only 1): sternum, costal p.: costal arch
O lumbar p.: consists of 2 crus:
O medial crus: median arcuate lig.: O, I: bo. LV 1-3 each right+left
O lateral crus: medial arcuate lig. (quadratus lumb. arcade): O: bo. LV1-2, I: cost. proc. LV1 as well as: lateral arcuate lig. (lateral arcuate arcade): O: cst. procc. LV1, I: 12th rib

Serratus posterior muscles
– Serratus posterior superior muscle
O spinous procc. CV6-7, TV1-2
F ribs: elevate
IN intercostal nerves (Th1-4)
– Serratus posterior inferior muscle
O thoracolumbal fascia, area of TV12 / LV1-3
I up to ribs 9-12
F ribs: depress
IN intercostal nerves (Th9-12)

Sternocleidomastoideus muscle
O sternum, clavicle
I mastoid proc.
F head: ROT, lateral bending, REC, clavicle: elevate = support of inspiration
IN accessory nerve, cervical pl. (C1-2)

Transversus thoracis muscle
O xiphoid proc. of sternum
I costal cartilages 2-6
F expiration
IN intercostal nerves 2-6

Intercostal musculature
IN intercostal nerves 1-11

External intercostal muscles
O superior border of rib
I superior border of rib
F ribs: elevate

Internal intercostal muscles
O superior border of rib
I inferior border of rib
F ribs: depress

Abdominal musculature
F of all: abdominal compression, discharge: upwards: forced expiration, coughing, downwards: urinate, defecation, giving birth, other than discharge: stabilization of trunk (lifting of weights)

External oblique
O ribs 5-12
I rectus sheath, inguinal lig., iliac crest
+F trunk: ROT, lateral bending (Th5- 12)
IN intercostal nerves (Th5- 12)

Internal oblique
O thoracolumb. fascia, inguinal lig., iliac crest
I lat. costal arch, rectus sheath
+F trunk: ROT, lateral bending
IN intercostal nerves (Th10- 12), L1

Transverse abdominal
O ribs 7-12, inguinal lig., iliac crest
I semilunar line of rectus sheath
IN intercostal nerves (Th7-12), L1

Rectus abdominis
O paramedian at the outer surface of costal arch, pubic bone, symphysis
I pubic bone
+F trunk: initiation of flexion (Th5- 12)
IN intercostal nerves (Th5- 12)
R 3-4 tendinous intersections

Pyramidal muscle
O pubic bone
I linea alba
F only: keeps linea alba taut
IN Th12, L1

Musculature of shoulder girdle

Muscles between trunk and shoulder girdle

Trapezius
O desc. p.: sup. nuchal line, nuchal lig, horiz. p.: CV7-TV3 asc. p.: TV3-TV12
I desc. p.: lat. third of clavicle, horiz. p.: acromion asc. p.: scapular spine
F scapula: maintain, ADD, ROT, participates in elevation of arm
IN accessory nerve, trapezius r. (C2-4)

Levator scapulae
O trv. procc. CV1-4
I superior scapular angle
F scapula: maintain, elevate, rotation of inf. angle towards med.
IN dorsal scapular nerve (C4- 5)

Rhomboid muscles
O spinous procc. CV6-7 and TV1-4
I med. margin of scapula
F scapula towards thorax and VC
IN dorsal scapular nerve (C4- 5)

Serratus anterior
O ribs 1-9
I med. margin of scapula
F scapula: maintain, elevate, rotation of inf. angle towards lat.
R is pre-requisite for elevation of arm; supports inspiration
IN long thoracic nerve (C5-7)

Pectoralis minor
O ribs 3-5
I coracoid proc.
F scapula: depress, ROT, support of inspiration
IN pectoral nerves (C6-8)

Subclavius
O ventr. cartilaginous part of 1st rib
I inferior surface of clavicle
F anchors and depresses clavicle
IN subclavian nerve (C5-6)

Muscles between trunk and upper arm

Latissimus dorsi
O TV7-12, lat. at ribs 10-12, posterior third of iliac crest, sacrum, spinous proc. LVC
I crest of lesser tubercle
F humeral joint: RV, ADD; shoulders: towards posteroinferior (forced expiration)
IN thoracodorsal nerve (C6 – 8)

Pectoralis major
O clavicular p.: med. at the clavicle, sternocostal p.: sternum, ribs 1-6 abdominal p.: anterior leaf of rectus sheath
I crest of greater tubercle C: p.: dist., s. p.: median, a. p.: prox.
F arm raised: depression, ADD, MR; arm abducted: C. p., s. p.: AV; accessory muscle of inspiration
IN pectoral nerves (C5-Th1)

Musculature of shoulder

Deltoid
O clavicular p., acromial p., spinal p. (scapular spine)
I deltoid tuberosity
F humeral joint: not: ADD when ABD>60° clavicular p.: AV, MR, ADD, ABD acromial p.: ABD spinal p.: RV, LR, ADD, ABD
IN axillary nerve (C4–6); clav. p.: + pectoral branches (C4–5)

Rotator cuff
Lateral rotators

Supraspinatus muscle
O supraspinatus fossa
I greater tubercle
F humeral joint: initiation of ABD, some LR
IN suprascapular nerve (C4–6)

Infraspinatus muscle
O infraspinatus fossa
I greater tubercle
F humeral joint: LR
IN suprascapular nerve (C4–6)

Teres minor muscle
O lat. margin of scapula
I greater tubercle
F humeral joint: some LR
IN axillary nerve (C5–6)

Medial rotators

Subscapularis muscle
O ventr. at the subscapular fossa
F humeral joint: MR
IN subscapular nerve (C5–8)

Teres major muscle
O inf. angle of scapula
F humeral joint: MR, ADD, RV
IN thoracodorsal nerve (C6–7)

Musculature of upper arm

Flexors

Biceps brachii muscle
O long h.: supraglenoid tubercle, short h.: coracoid proc.
I tuberosity of radius, ulnar at fascia of forearm (lacertus fibrosus)
F humeral joint: AV, long h.: ABD, MR; short h.: ADD; cubital joint: FLX, SUP
IN musculocutaneous nerve (C5–6)

Coracobrachialis muscle
O coracoid proc.
I dist. to crest of lesser tubercle
F humeral joint: AV
IN musculocutaneous nerve (C6–7)

Brachialis muscle
O dist. half of anterior surface of humerus
I ulnar tuberosity
F cubital joint: FLX
IN musculocutaneous nerve (C5–6)

Extensors

Triceps brachii muscle
O long h.: infraglenoid tubercle of scapula; med. h.: dist. dorsal surface of humerus, lat. h.: dist. to the groove for radial nerve
I olecranon
F humeral joint: RV; cubital joint: EXT
IN radial nerve (C6–8)

Musculature of forearm

Flexors
Superficial layer
O med. epicondyle of humerus

Pronator teres muscle
O humeral h.: med. epicondyle of humerus; ulnar h.: ulnar coronoid proc.
I tuberosity of radius
F PRO, cubital joint: FLX
IN median nerve (C6–7)

Flexor carpi radialis muscle
I base of 2nd metacarpal bone
F cubital joint: some FLX, PRO; hand: pFLX, rABD
IN median nerve (C6–7)

Palmaris longus muscle
I palmar aponeurosis
F wrist joints: FLX
IN median nerve (C7–Th1)
R may be absent

Flexor carpi ulnaris muscle
O humeral h.: med. epicond. of humerus, ulnar h.: olecranon, prox. ulna
I pisiform bone, 5th metacarpal bone, hook of the hamate
F wrist joints: pFLX, uABD
IN ulnar nerve (C7–8)

Intermediate layer

Flexor digitorum superficialis muscle
O humeral/ulnar h.: med. epicond. of humerus, coronoid proc. of ulna, radial h.: palmar radius
I med. phalanges
F cubital joint; some FLX, wrist joints/PIP: FLX
IN median nerve (C7–Th1)
R tendon gaps for passage of the tendons of profundus muscle

Deep layer

Flexor pollicis longus muscle
O radius (palmar), interosseous membrane
I dist. phalanx of thumb
F thumb: FLX, rABD
IN median nerve (C7–8)

Flexor digitorum profundus muscle
O ulna (palmar), interosseous membrane
I dist. 2nd to 5th phalanges
F wrist joints, interphalangeal joints: FLX
IN median and ulnar nerve (C7–Th1)
R tendons pass through tendons of superficialis muscle

Pronator quadratus muscle
O dist. palmar at ulna
I dist. palmar at radius
F PRO
IN median nerve (C8–Th1)

Extensors
Superficial (ulnar) layer
O caput commune: lat. epicondyle, lat. collat. lig., annular lig. of radius, ulna

Extensor digitorum communis muscle
I dorsal aponeurosis of 2nd to 5th
F wrist joints: dEXT, uABD, digits: EXT, spreading
IN radial nerve, deep ra. (C6–8)
R its tendons are connected via 3 intertendinous connections

Extensor digiti minimi muscle
I dorsal aponeurosis of 5th
F wrist joint: dEXT, uABD, little finger: EXT
IN radial nerve, deep ra. (C6–8)

Extensor carpi ulnaris muscle
I base of 5th metacarpal
F wrist joints: only uABD!
IN radial nerve, deep ra. (C7–8)

Deep (radial) layer
O sequence of muscles is according to their origins from prox. to dist.; dors.: ulna, interosseous membrane, radius

Supinator muscle
O supinator crest of ulna, caput commune (see above)
I ant. surface of radius (palmar)
F SUP
IN radial nerve, deep ra. (C5–6)

Abductor pollicis longus muscle
I base of 1st metacarpal
F thumb: ABD, wrist joints: pFLX, rABD
IN radial nerve, deep ra. (C7–8)

Extensor pollicis brevis muscle
O base of 1st prox. phalanx
I thumb, wrist joint: EXT, rABD
IN radial nerve, deep ra. (C7-Th1)

Extensor pollicis longus muscle
O middle dorsal surface of ulna
I base of distal phalanx of thumb
F thumb: EXT, wrist joint: dEXT, rABD
IN radial nerve, deep ra. (C7- 8)

Extensor indicis muscle
O dist. dorsal surface of ulna, interosseous membrane
I dorsal aponeurosis of 2nd finger
F index: EXT, wrist joint: dEXT
IN radial nerve, deep ra. (C6-8)

Radial group

Brachioradialis muscle
O lat. supracondylar crest of humerus, lat. intermuscular septum
I styloid proc. of radius
F cubital joint: FLX, depending on position PRO or SUP
IN radial nerve (C5-6)

Extensor carpi radialis longus muscle
O lat. supracondylar crest of humerus, lat. intermuscular septum
I base of 2nd metacarpal
F cubital joint: Some FLX, arm flexed: PRO, arm extended: SUP, hand: dEXT, rABD
IN radial nerve, deep ra. (C6- 7)

Extensor carpi radialis brevis muscle
O caput commune (see ulnar extensors)
I base of 3rd metacarpal
F cubital joint: some FLX, wrist joint: uABD; dEXT
IN radial nerve, deep ra. (C7)

Short muscles of hand

Palm
O dorsal aponeurosis
F MP: FLX; PIP and DIP: EXT

Lumbricales muscles (total of 4)
O radial border of the tendons of the flexor digitorum profundus muscle
I rad. at dorsal aponeurosis
IN lumbr. m. 2-3: median nerve, lumbr. m. 4-5: ulnar nerve, deep ra. (C8-Th1)

Dorsal interossei muscles (total of 4)
O two-headed at each of 2 metacarpal bones
I dorsal aponeuroses; index: rad.; dig. 3: rad. & uln.; dig. 4: uln.
F ABD
IN ulnar nerve, deep ra. (C8-Th1)

Palmar interossei muscles (total of 3)
O one-headed at 2nd, 4th, 5th metacarpal bone
I base of 2nd, 4th, 5th prox. phalanges, dorsal aponeur.
F ADD
IN ulnar nerve, deep ra. (C8-Th1)

Thenar area
O flexor retinaculum
I prox. phalanx (except opponens muscle)

Abductor pollicis brevis muscle
+O tubercle of scaphoid
+I via rad. sesamoid bone
F thumb: pABD
IN median nerve (C8-Th1)

Opponens pollicis muscle
+O rad. border of 1st metacarpal bone
F thumb: OPP, ADD
IN median nerve (C6-7)

Flexor pollicis brevis muscle
+O superf. h.: flexor retinaculum; deep h.: trapezium bone, trapezoid bone, capitate bone
+I via rad. sesamoid bone
F thumb: OPP; MP: FLX, ADD
IN superf. h.: median nerve, deep h.: ulnar nerve (C8- Th1)

Adductor pollicis muscle
+O transverse h.: entire length of 3rd metacrp. bone oblique h.: capitate bone, trapezoid bone
+I via uln. sesamoid bone
F thumb: ADD, OPP, MP: FLX
IN ulnar nerve, deep ra. (C8- Th1)

Hypothenar area
IN ulnar nerve, deep ra. (C8-Th1), (except palmaris brevis muscle)

Palmaris brevis muscle
O palmar aponeurosis, flexor retinaculum
I skin of hypothenar area
F formation of palm
IN ulnar nerve, superf. ra. (C8-Th1)

Abductor digiti minimi muscle
O pisiform bone, flexor retinaculum
I base of prox. phalanx of 5th digit
F small finger: only ABD

Flexor digiti minimi brevis muscle
O flexor retinaculum, hook of hamate
I base of prox. phalanx of 5th digit
F small finger: FLX
R often missing

Opponens digiti minimi muscle
O flexor retinaculum, hook of hamate
I outer border of 5th metacarpal bone
F small finger: OPP

Musculature of hips

Extensors

Gluteus maximus muscle
O iliac crest, sacrum, sacrotuberal lig.
I prox. portion: iliotibial tract; dist. portion: gluteal tuberosity
F hip joint: prox. portion: EXT, LR, ABD; dist. portion: EXT, LR, ADD
IN inf. gluteal nerve (L5-S2)

Flexors

Iliopsoas muscle
I lesser trochanter
F hip joint: FLX, LR, INC, LVC: lateral bending
R antagonist of erector spinae muscle

– Psoas major muscle
O superficial portion: vertebral bodies TV12, LV1-4; deeper portion: costal proce. LVC
IN lumbar pl., femoral nerve (L1-3)

– Iliacus muscle
O iliac fossa
IN lumbar pl., femoral nerve (L2-4)

Abductors

Gluteus medius/minimus muscle
O gluteal surface of wing of ilium
I greater trochanter (tip)
F hip joint: ventral portion: FLX, MR; dorsal portion: EXT, LR
IN sup. gluteal nerve (L4-S1)

Tensor fascia lata muscle
O ant. sup. iliac spine iliotibial tract
F hip joint: FLX, MR, ABD; iliotibial tract: tense
IN sup. gluteal nerve (L4-5)

Adductors

Pectineus muscle
O pecten of pubis
I pectineal line (at lesser trochanter)
F hip joint: ADD, FLX, some MR
IN femoral nerve (L2-3), obturator nerve, ant. ra. (L2-4)

Adductor longus / brevis muscle
R brevis below longus
O sup. ra. of pubis (close to symphysis)
I linea aspera
F hip joint: ADD, LR, some FLX
IN obturator nerve, ant. ra. (L2-4)

Gracilis muscle
O inf. ra. of pubis (close to symphysis)
I superf. pes anserinus
F hip joint: ADD, LR; knee joint: FLX, MR; safeguarding of med. knee joint
IN obturator nerve, ant. ra. (L2-4)

Adductor magnus/minimus muscle
O inf. ra. of pubis, ra. of ischium, tuberosity of ischium
I linea aspera, med. epicond.
F hip joint: ADD, EXT
IN obturator nerve, ant. ra. (L2-4), tibial nerve (L3-5)

Lateral rotators

Piriformis muscle
O pelvic surface of sacrum (inner side)
I greater trochanter
F hip joint: LR, ABD
IN sacral pl. (L5-S2)

Obturator internus muscle
O hip bone near obturator for., obturator membrane
I trochanteric fossa
F hip joint: LR; ABD in seated position
IN obturator nerve (L1-4)

Obturator externus muscle
O hip bone near obturator for., obturator membrane
I trochanteric fossa
F hip joint: LR, some ABD
IN obturator nerve (L1-4)

Gemelli muscles:
R above and below the int. obturator muscle
I trochanteric fossa
F hip joint: LR in seated position: ABD
IN inf. gluteal nerve, sacral pl. (L5-S2)
- Gemelli superior muscle
O ischial spine
- Gemelli inferior muscle
O ischial tuberosity

Quadratus femoris muscle
O ischial tuberosity
I intertrochanteric crest
F hip joint: LR, ADD
IN inf. gluteal nerve, sacral pl. (L5-S2)

Musculature of upper thigh

Extensor group

Quadriceps femoris muscle:
I via patella, patellar lig. at the tibial tuberosity
F knee joint: EXT
IN femoral nerve (L2-4)
- Rectus femoris muscle
O ant. inf. iliac spine
F knee joint: EXT; hip joint: FLX
- Vastus medialis muscle
O linea aspera
- Vastus intermedius muscle
O anterior and lateral surface of femur
- Vastus lateralis muscle
O greater trochanter up to the linea aspera

Sartorius muscle
O ant. sup. iliac spine
I superf. pes anserinus
F hip joint: FLX, LR, knee joint: FLX
IN femoral nerve (L1-3)

Flexor group
= Ischiocrural musculature

Semitendinosus muscle
O tuberosity of ischium
I superf. pes anserinus
F hip joint: EXT; knee joint: FLX, MR
IN tibial nerve (L5-S2)

Semimembranosus muscle
O tuberosity of ischium
I med. condyle of tibia, fascia, popliteus muscle, as oblique popliteal lig. into the capsule
F combined these insertions are also called the deep pes anserinus
F hip joint: EXT; knee joint: FLX, MR
IN tibial nerve (L5-S2)

Biceps femoris muscle
O long h.: tuberosity of ischium, short h.: linea aspera
O head of fibula
F hip joint: EXT; knee joint: FLX, LR, lat. safeguarding of knee joint
IN long h.: tibial nerve (L5-S2) short h.: common peroneal nerve (S1-S2)

Musculature of lower thigh

Extensor group

Tibialis anterior muscle
O lat. surface of tibia, interosseous membrane
I med. cuneiform, 1st metatarsal bone
F UAJ: dEXT, LAJ: pFLX, SUP, ADD
IN deep peroneal nerve (L4-5)

Extensor digitorum longus muscle
O lat. cond. of tibia, interosseous membrane, head of fibula
I dorsal aponeuroses of 2nd to 5th toes
F foot, digits: dEXT
IN deep peroneal nerve (L5-S1)

Extensor hallucis longus muscle
O middle of fibula, interosseous membrane
I dist. phalanx of great toe
F foot, great toe: dEXT
IN deep peroneal nerve (L4-S1)

Deep flexors

Tibialis posterior muscle
O interosseous membrane, partly tibia, fibula
I navicular tuberosity, cuneiform bone
R embraces from below the calcaneonavicular lig.
F foot: pFLX, SUP
IN tibial nerve (L4-5)

Flexor digitorum longus muscle
O post. surface of tibia
I dist. phalanges of 2nd to 5th digit
F foot: pFLX, SUP; digits: pFLX
R longitudinal bracing of arch of foot
IN tibial nerve (S1-3)

Flexor hallucis longus muscle
O dist. posterior surface of fibula, interosseous membrane
I dist. phalanx of great toe
R bracing of foot
F great toe: pFLX
IN tibial nerve (S1-3)

Superficial flexors

Triceps surae muscle
I tuberosity of calcaneus via calcaneal tendon
F foot: pFLX
IN tibial nerve (S1-2)
- Gastrocnemius muscle
O med. h.: med. epicond. of femur, lat. h.: lat. epicond. femur
- Soleus muscle
O head of fibula, soleal line of tibia

Plantaris muscle
- O lat. cond. of femur
- I calcaneal tendon
- F foot; dEXT
- IN tibial nerve (S1-2)

Popliteus muscle
- O lat. epicond. of femur
- I post. surface of tibia
- F knee joint: FLX, MR
- IN tibial nerve (L4-S1)

Peroneus group
- IN superf. peroneal nerve (L5-S1)
- R longus covers brevis

Peroneus longus muscle
- O prox. at fibula
- I tuberosity of 1st metatarsal bone, med. cuneiform bone
- F foot: PRO, pFLX

Peroneus brevis muscle
- O lat. at fibula
- I tuberosity of 5th metatarsal
- F foot: PRO, pFLX

Musculature of foot

Muscles of back of foot
- I dorsal at calcaneus
- IN deep peroneal nerve (S1-2)

Extensor digitorum brevis muscle
- I dorsal aponeuroses of 2nd to 4th digit
- F these toes: dEXT

Extensor hallucis brevis muscle
- I dorsal aponeuroses of great toe
- F great toe: dEXT

Muscles of sole of foot

Thenar area

Abductor hallucis muscle
- O tuberosity of calcaneus
- I med. sesamoid bone, base of prox. phalanx of 1st digit
- F great toe: ABD; some FLX; maintains arch
- IN med. plantar nerve (L5-S1)

Flexor hallucis brevis muscle
- O med. cuneiform bone
- I sesamoid bone of prox. phalanx of 1st digit
- F great toe: pFLX
- IN med. plantar nerve (L5- S1)

Adductor hallucis muscle
- O oblique h.: cuboid bone, lat. cuneiform bone, bases of metatarsals of 2nd and 3rd digit, transverse h.: metatarsophalangeal joints of 3rd to 5th digit
- I via lat. sesamoid bone of prox. phalanx of 1st digit
- F great toe: FLX, ADD; maintains arch
- IN lat. plantar nerve, deep ra. (S1-S2)

Hypothenar area
- IN lat. plantar nerve (S1-2)

Abductor digiti minimi muscle
- O lat. proc. of tuberosity of calcaneus
- I prox. phalanx of 5th digit
- F small toe: pFLX, some ABD; maintains arch

Flexor digiti minimi muscle
- O base of metatarsal of 5th digit
- I base of metatarsal of 5th digit
- F small toe: pFLX

Opponens digiti minimi muscle
- O tendon sheath of peroneus longus muscle
- I base of metatarsal of 5th digit
- F moves the metatarsal of 5th digit in a plantar direction
- R often missing

Middle area of sole of foot

Flexor digitorum brevis muscle
- O plantar at tuberosity of calcaneus
- I med. phalanx of 2nd to 4th digit
- F digits: pFLX
- IN med. plantar nerve (L5-S1)
- R tendon gaps for passage of longus

Quadratus plantae muscle
- O two heads med. / lat. at calcaneus
- I from lat. at tendons of flexor digitorum longus muscle
- F corrects direction of pull of flexor digitorum longus muscle
- IN lat. plantar nerve (S1-2)

Lumbricales muscles (total of 4)
- O med. from tendons of flexor digitorum longus muscle
- I dorsal aponeuroses of 2nd to 5th digit
- F digits: MP: FLX; ADD towards great toe
- IN med. plantar nerve to lumbrical m. No. 2-4, lat. plantar n. to lumbricalis muscle No. 5

Interossei plantares muscles (total of 3)
- O medial surface of metatarsals of 3rd to 5th digit
- I medial side of base of prox. phalanx of 3rd to 5th digit
- F ADD digits 3 to 5 towards digit 2; pFLX
- IN lat. plantar nerve, deep ra. (S1-2)

Interossei dorsales muscles (total of 4)
- O adjacent sides of all metatarsals
- I base of prox. phalanges: metat. bone 1 and 2: me. and lat. at digit 2, m. bone 3: lat. at digit 3; m. bone 4: lat. at digit 4
- F digits 2-4: ABD
- IN lat. plantar nerve, deep ra. (S1-2)

Abbreviations

- ABD abduction
- ADD adduction
- ant. anterior
- asc. ascending
- AV antevarum
- brev. brevis (short)
- bo. body, bodies
- CV(C) cervical vertebra(l) (column)
- Cond. condyle
- Cst. proc. costal process
- dEXT dorsal extension
- Dig. digit, digits
- DIP distal interphalangeal joint
- dist. distal
- dors. dorsal, dorsally or alike
- EXT extension
- F (+F) function(s), (addendum toF)
- FLX flexion
- For. foramen
- H. head
- horiz. horizontal
- interphalang. interphalangeal
- I (+I) insertion (addendum to I)
- IN innervation
- INC inclination
- inf. inferior
- LAJ lower ankle joint
- LR lateral rotation
- lat. lateral, laterally or alike
- Lig. ligament
- long. longus (long)
- M. muscle
- med. medial or alike
- MP metacarpophalangeal joint
- MR medial rotation
- N. nerve
- OPP opposition
- O (+O) origin (addendum to O)
- P. part
- pFLX plantar / plantar flexion
- pABD palmar abduction
- PIP proximal interphalangeal joint
- Pl. plexus
- post. posterior
- PRO pronation
- Proc(c). process(es)
- R remark
- Ra. ramus, branch
- rABD radial abduction
- Rad radial
- REC reclination
- ROT rotation
- RV retroversion
- sup. superior
- Spn. proc. spinal process
- SUP supination
- superf. superficial
- temp. temporal
- Trv. proc. transverse process
- prox. proximal or alike
- TV(C) thoracic vertebra(l) (column)
- UAJ upper ankle joint
- Uln ulnar
- uABD ulnar abduction
- ventr. ventral
- VC vertebral column

Clinically Relevant Muscles and their Innervation

Upper Extremities (brachial plexus C5–T1)			
Muscle	Myotomes	Nerves	Test
Supra- and infraspinatus	C4–C6	suprascapul. nerve	lateral rotation of the upper arm
Serratus anterior	C5–C7	long thoracic nerve	forward elevation of the arm
Pectoralis major	C5–T1	pectoral nerves	joining hands in front of the body
Deltoid	C5–C6	axillary nerve	abduction of the upper arm
Biceps and brachialis	C5–C6	musculocut. nerve	flexion of the arm in supination
Brachioradialis	C5–C6	radial nerve	flexion of the arm in mid position
Triceps brachii	C6–C8	radial nerve	extension of the arm
Extensor carpi radialis & ulnaris	C6–C8	radial nerve	dorsiflexion of the hand
Extensor digitorum communis	C6–C8	radial nerve	finger extension in all joints
Extensor pollic., abduct. pollic. long.	C7–C8	radial nerve	extension and abduction of the thumb
Pronator teres	C6–C7	median nerve	pronation of the flexed arm
Flexor carpi radialis	C6–C7	median nerve	palmar flexion and radial abduction
Flexor digitorum superfic.	C7–T1	median nerve	flexion of fingers in proximal/mid joint
Flexor pollicis & digitorum II, III, profundus	C7–C8	median nerve	flexion of 1st–3rd fingers in distal joint
Abductor pollicis brevis	C7–C8	median nerve	insufficient abduction in grip test
Opponens pollicis	C7/C8	median nerve	thumb-pinky opposition
Flexor carpi ulnaris	C7–T1	ulnar nerve	palmar flexion and ulnar abduction
Flexor digitalis IV, V prof.	C7–T1	ulnar nerve	flexion of 5th finger in distal joint
Interossei	C8–T1	ulnar nerve	spreading and adduction of fingers
Adductor pollicis	C8–T1	ulnar nerve	adduction of the thumb (Froment's-sign)
Hypothenar (abductor, oppon., flex. brev. digiti quinti)	C8–T1	ulnar nerve	flexion, opposition and abduction of the little finger in the proximal joint
Lumbricales	C7–T1	median nerve (dig. I–II) ulnar nerve (dig. III–IV)	flexion of the fingers in the prox. joints and extension in the mid and dist. joints

Lower Extremities (lumbosacral plexus L1–S3)			
Muscle	Myotomes	Nerves	Test
Iliopsoas	L1–L4	femoral nerve	hip flexion
Quadriceps femoris	L2–L4	femoral nerve	knee extension
Adductor longus, magnus, brevis and gracilis	L2–L4	obturator nerve	hip adduction
Gluteus medius and minimus	L4–S1	superior gluteal nerve	hip abduction (Trendelenburg sign)
Gluteus maximus	L5–S2	inferior gluteal nerve	hip extension
Biceps femoris	L5–S2	sciatic (cap.long.) nerve com. peron. (c.brev.) n.	knee flexion
Semitend. and semimembran.	L4–S2	sciatic nerve	knee flexion
Tibialis anterior	L4–L5	deep peron. nerve	dorsiflexion of the foot
Extensor hallucis longus	L5	deep peron. nerve	hallux extension
Extensor digitorum longus	L5–S1	deep peron. nerve	toe extension (toes II–V)
Peronei	L5–S1	superf.peron.nerve	eversion of the foot
Extensor digitorum brevis	L5–S1	deep peron. nerve	toe extension

Cervical Plexus

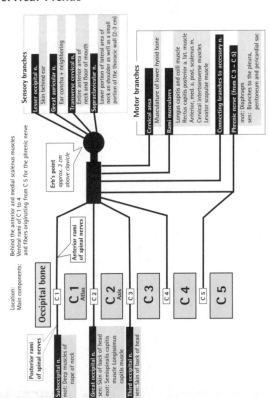

Sensory branches

Lesser occipital n.
Skin behind ear

Great auricular n.
Ear concha + neighboring

Transverse cervical n.
Entire anterior area of neck and floor of mouth

Supraclavicular n.
Lower portion of lateral area of neck an shoulder as well as a small portion of the thoracic wall (2–3 cm)

Motor branches

Cervical ansa
Musculature of lower hyoid bone

Rami musculares
Longus capitis and colli muscle
Rectus capitis posterior a. lat. muscle
Anterior, med. a. post. scalenus m.
Cervical intertransverse muscles
Levator scapulae muscle

Connecting branches to accessory n.

Phrenic nerve (from C 3 – C 5)
mot: Diaphragm
sen: Branches to the pleura, peritoneum and pericardial sac

Erb's point
approx. 3 cm
above clavicle

Location: Behind the anterior and medial scalenus muscles
Main components: Ventral rami of C 1 to 4
and fibers originating from C 5 for the phrenic nerve

Anterior rami
of spinal nerves

Occipital bone

C 1 Atlas

C 2 Axis

C 3

C 4

C 5

Posterior rami
of spinal nerves

Suboccipital n.
mot: Deep muscles of nape of neck

Great occipital n.
sen: Skin of back of head
mot: Semispinalis capitis muscle Longissimus capitis muscle

Third occipital n.
sen: Skin of back of head

Brachial Plexus

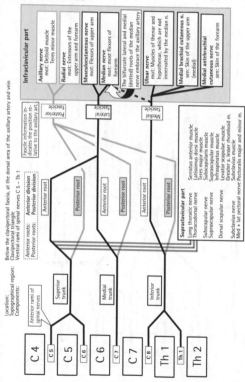

Infraclavicular part

Axillary nerve
mot: Deltoid muscle
Teres minor muscle

Radial nerve
mot: Extensors of the upper arm and forearm

Musculocutaneous nerve
mot: Flexors of upper arm

Median nerve
mot: most flexors of forearm

The bifurcate lateral and medial (dotted) roots of the median nerve embrace the axillary artery.

Ulnar nerve
mot: Muscles of thenar and hypothenar, which are not innervated by the median n.

Medial brachial cutaneous n.
sen: Skin of the upper arm (medial)

Medial antebrachial cutaneous n.
sen: Skin of the forearm

Fascicle information indicates the position relative to the axillary art.

Posterior fascicle
Lateral fascicle
Medial fascicle

Below the claviphpectoral fascia, at the dorsal area of the axillary artery and vein
Clavipectoral triangle
Ventral rami of spinal nerves C 5 – Th 1

Anterior roots: Anterior division
Posterior roots: Posterior division

Anterior root
Posterior root
Anterior root
Posterior root
Anterior root
Posterior root

Supraclavicular part
Long thoracic nerve   Serratus anterior muscle
Thoracodorsal nerve   Latissimus dorsi muscle
                             Teres major muscle
Subscapular nerve   Subscapularis muscle
Suprascapular nerve   Supraspinatus muscle
Dorsal scapular nerve   Infraspinatus muscle
                             Levator scapulae muscle
                             Greater a. lesser rhomboid m.
Subclavius nerve   Subclavius muscle
Med + lat pectoral nerve   Pectoralis major and minor m.

Location:
Topographical region:
Components:

Anterior rami of spinal nerves

Superior trunk
Medial trunk
Inferior trunk

C 4
C 5
C 6
C 7
C 8
Th 1
Th 2

Lumbar Plexus

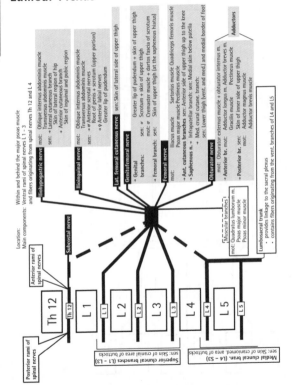

Location: Within and behind the major psoas muscle

Main components: Ventral rami of spinal nerves L1–3 and fibers originating from spinal nerves Th12 and L4

Iliohypogastric nerve: mot: Oblique internus abdominis muscle
Transversus abdominis muscle
→ Lateral cutaneous branch sen: Skin of lateral region of hip
→ Anterior cutaneous branch sen: Skin of inguinal and pubic region

Ilioinguinal nerve: mot: Oblique internus abdominis muscle
Transversus abdominis muscle
→ Anterior scrotal nerves sen: Root of penis + scrotum (upper portion)
→ Anterior labial nerves Greater lip of pudendum

Lat. femoral cutaneous nerve sen: Skin of lateral side of upper thigh

Genitofemoral nerve:
Genital branches: sen: ♀ Greater lip of pudendum + skin of upper thigh
mot: ♂ Scrotum + skin of upper thigh
mot: ♂ Cremaster muscle + dartos fascia of scrotum
Femoral branch: sen: Skin of upper thigh (at the saphenous hiatus)

Femoral nerve: mot: Iliacus muscle Sartorius muscle Quadriceps femoris muscle
Psoas major muscle Pectineus muscle
→ Ant. cutaneous branches sen: Anterior side of upper thigh up to the knee
→ Saphenous n. → Infrapatellar branch: sen: Medial skin below patella
→ Med. crural cutan. branch:
sen: Lower thigh (vent. and med.) and medial border of foot

Obturator nerve: mot: Obturator externus muscle + obturator internus m.
→ Anterior br. mot: Adductor longus m. Adductor brevis m.
Gracilis muscle Pectineus muscle
sen: Skin of inner side of upper thigh
→ Posterior br. mot: Adductor magnus muscle
Adductor brevis muscle

Adductors

Lumbosacral trunk
· provides linkage to the sacral plexus
· contains fibers originating from the vent. branches of L4 and L5

Anterior rami of spinal nerves

Subcostal nerve

Th12 L1 L2 L3 L4 L5

Muscular branches
mot: Quadratus lumborum m.
Psoas major muscle
Psoas minor muscle

Posterior rami of spinal nerves

Superior cluneal branches (L1–L3) sen: Skin of cranial area of buttocks

Medial cluneal bran. (L4–S3) sen: Skin of cranimed, area of buttocks

Sacral Plexus

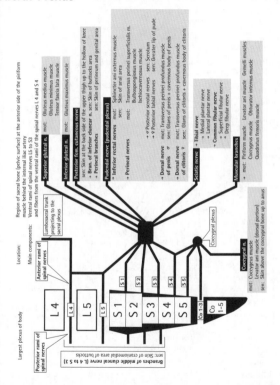

Location: Region of sacral bone (pelvic surface) at the anterior side of the piriform muscle behind the internal iliac artery

Main components: Ventral rami of spinal nerves L5 to S3 and fibers from the ventral rami of the spinal nerves L 4 and S 4

Superior gluteal n. mot: Gluteus medius muscle
Gluteus minimus muscle
Tensor fascia lata muscle

Inferior gluteal n. mot: Gluteus maximus muscle

Posterior fem. cutaneous nerve
sen: Skin at the back side of the upper thigh up to the hollow of knee
→ **Bran. of inferior cluneal n.** sen: Skin of buttocks area
→ **Perineal branches** sen: Skin of perineum and genital area

Pudendal nerve (pudendal plexus)
→ **Inferior rectal nerves** mot: Sphincter ani externus muscle
sen: Skin of anal area
→ **Perineal nerve** mot: Transversus perinei superficialis m.
Bulbospongiosus muscle
Ischiocavernosus muscle
→ ♂ Posterior scrotal nerves sen: Scrotum
→ ♀ Posterior labial nerves sen: Greater lip of pude
→ **Dorsal nerve of penis** mot: Transversus perinei profundus muscle
sen: Glans of penis + cavernous body of penis
→ **Dorsal nerve of clitoris** ♀ mot: Transversus perinei profundus muscle
sen: Glans of clitoris + cavernous body of clitoris

Sciatic nerve
→ **Tibial nerve**
→ Medial plantar nerve
→ Lateral plantar nerve
→ **Common fibular nerve**
→ Superficial fibular nerve
→ Deep fibular nerve

Muscular branches
mot: Piriform muscle   Levator ani muscle   Gemelli muscles
Coccygeus muscle   Obturator internus muscle
Quadratus femoris muscle

Coccygeal n. (dorsal portion)
mot: Coccygeus muscle
Levator ani muscle
sen: Skin above the coccygeal bone up to the anus

Anterior rami of spinal nerves

Lumbosacral trunk projecting to the sacral plexus

Coccygeal plexus

L 4
L 5
S 1
S 2
S 3
S 4
S 5
Co 1-3
Co 1-5

Largest plexus of body

Posterior rami of spinal nerves

sen: Skin of craniomedial area of buttocks
Branches of middle cluneal nerve (L 4 to S 3)

Cutaneous Innervation of the Body

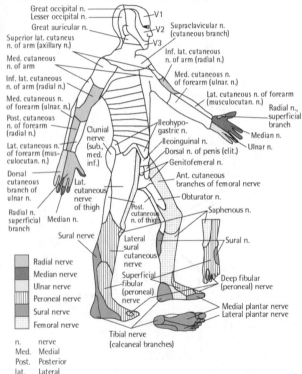

Great occipital n.
Lesser occipital n.
Great auricular n.
Superior lat. cutaneus n. of arm (axillary n.)
Med. cutaneous n. of arm
Inf. lat. cutaneous n. of arm (radial n.)
Med. cutaneous n. of forearm (ulnar n.)
Post. cutaneous n. of forearm (radial n.)
Lat. cutaneous n. of forearm (musculocutan.)
Dorsal cutaneous branch of ulnar n.
Radial n. superficial branch
Median n.
Sural nerve

V1
V2
V3
Supraclavicular n. (cutaneous branch)
Inf. lat. cutaneous n. of arm (radial n.)
Med. cutaneous n. of forearm (ulnar. n.)
Lat. cutaneous n. of forearm (musculocutan.)
Radial n., superficial branch
Median n.
Ulnar n.

Clunial nerve (sub., med. inf.)
Ileohypogastric n.
Ileoinguinal n.
Dorsal n. of penis (clit.)
Genitofemoral n.
Ant. cutaneous branches of femoral nerve
Obturator n.
Saphenous n.
Sural n.
Deep fibular (peroneal) nerve
Medial plantar nerve
Lateral plantar nerve

Lat. cutaneous nerve of thigh
Post. cutaneous n. of thigh
Lateral sural cutaneous nerve
Superficial fibular (peroneal) nerve
Tibial nerve (calcaneal branches)

Radial nerve
Median nerve
Ulnar nerve
Peroneal nerve
Sural nerve
Femoral nerve

n.   nerve
Med.   Medial
Post.   Posterior
lat.   Lateral

The Dermatome Map

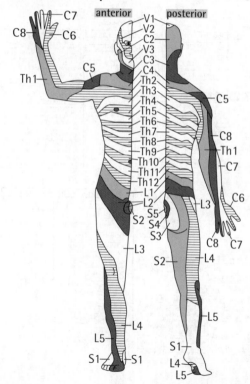

Abdominal Aorta

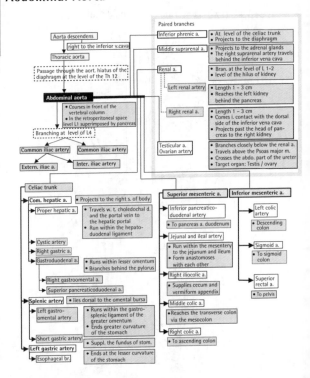

Paired branches

Inferior phrenic. a.
- At level of the celiac trunk
- Projects to the diaphragm

Middle suprarenal a.
- Projects to the adrenal glands
- The right suprarenal artery travels behind the inferior vena cava

Renal a.
- Bran. at the level of L 1–2
- level of the hilus of kidney

Left renal artery
- Length 1 – 3 cm
- Reaches the left kidney behind the pancreas

Right renal a.
- Length 1 – 3 cm
- Comes i. contact with the dorsal side of the inferior vena cava
- Projects past the head of pancreas to the right kidney

Testicular a. Ovarian artery
- Branches closely below the renal a.
- Travels above the Psoas major m.
- Crosses the abdo. part of the ureter
- Target organ: Testis / ovary

Aorta descendens
- right to the inferior v.cava

Thoracic aorta

Passage through the aort. hiatus of the diaphragm at the level of the Th 12

Abdominal aorta
- Courses in front of the vertebral column
- In the retroperitoneal space level L1 superimposed by pancreas

Branching at level of L4

Common iliac artery **Common iliac artery**

Extern. iliac a. **Inter. iliac artery**

Celiac trunk

Com. hepatic a. — Projects to the right s. of body

Proper hepatic a.
- Travels w. t. choledochal d. and the portal vein to the hepatic portal
- Run within the hepato-duodenal ligament

Cystic artery

Right gastric a.

Gastroduodenal a.
- Runs within lesser omentum
- Branches behind the pylorus

Right gastroomental a.

Superior pancreaticoduodenal a.

Splenic artery — lies dorsal to the omental bursa

Left gastro-omental artery
- Runs within the gastro-splenic ligament of the greater omentum
- Ends greater curvature of the stomach

Short gastric artery — Suppl. the fundus of stom.

Left gastric artery

Esophageal br. — Ends at the lesser curvature of the stomach

Superior mesenteric a.

Inferior pancreatico-duodenal artery
- To pancreas a. duodenum

Jejunal and ileal artery
- Run within the mesentery to the jejunum and ileum
- Form anastomoses with each other

Right iliocolic a.
- Supplies cecum and vermiform appendix

Middle colic a.
- Reaches the transverse colon via the mesocolon

Right colic a.
- To ascending colon

Inferior mesenteric a.

Left colic artery
- Descending colon

Sigmoid a.
- To sigmoid colon

Superior rectal a.
- To pelvis

Mnemonics

Vessels	
Branches of internal iliac artery (1)	
I	Ileolumbar artery
Love	Lateral sacral arteries
Going	Gluteal artery (superior and inferior)
Places	Pudendal artery (internal)
In	Inferior vesical artery (lower part of bladder)
My	Middle rectal artery
Very	Vaginal artery
Own	Obturator artery
Underwear	Umbilical artery
Branches of internal iliac artery (2)	
Anterior	
I	Inferior gluteal artery
Milked	Middle rectal artery
Our	Obturator artery
Insatiable	Inferior vesical artery
Intern's	Internal pudendal artery
Udders	Umbilical artery
Under the Desk	U/D=Uterine a. (female)/ Deferential a. (male)
Posterior	
Put	Posterior branch
In	Iliolumbar artery
Left	Lateral sacral artery
Sock	Superior gluteal artery
Branches of external carotid artery	
Suzy	Superior thyroid artery
Always	Ascending pharyngeal artery
Lays	Lingual artery
Flat	Facial artery
On	Occipital artery
Pillows	Posterior auricular artery
Making	Maxillary artery
Sex Terrific	Superficial Temporal artery
Main branches of subclavian artery	
Vampire	Vertebral artery
In	Internal thoracic artery
The	Thyrocervical trunk
Car	Costocervical trunk
Terry!	Transverse cervical artery

Bones	
Cranial bones	
Old	Occipital
Pygmies	Parietal
From	Frontal
Thailand	Temporal
Eat	Ethmoid
Skulls	Sphenoid
Carpal bones	
Some	Scaphoid
Lovers	Lunate
Try	Triquetrium
Positions	Pisiform
That	Trapezium
They	Trapezoid
Cannot	Capate
Handle	Hamate
Tarsal bones	
Tall	Talus
Californian	Calcaneus
Navy	Navicular
Medical	Medial cuneiform
Interns	Intermediate cuneiform
Lay	Lateral cuneiform
Cuties	Cuboid
Number of vertebrae	
Breakfast at 7 AM	CVC: 7
Lunch at 12 PM	TVC: 12
Dinner at 5 PM	LVC: 5
Rotational movements of the forearm	
Palm to the sun	SUPINATION
Palm to the plants	PRONATION

Nerves and Plexuses

Cranial nerves

On	1. Olfactory nerves
Old	2. Optic nerve
Olympus	3. Oculomotor nerve
Towering	4. Trochlear nerve
Top	5. Trigeminal nerve
A	6. Abducent nerve
Famous	7. Facial nerve
Vocal	8. Vestibulocochlear nerve
German	9. Glossopharyngeal nerve
Viewed	10. Vagus nerve
Some	11. Spinal accessory nerve
Hops	12. Hypoglossal nerve

Modalities of nerve fibers of cranial nerves

Some Say Money Matters But My Brother Says Big Boobs Make More Sense

From I–XII, M=motor, S=sensory, B=both

Brachial plexus (branches and organization)

RAP: Radial nerve and axillary nerve originate from posterior fascicle

LAMM: Lateral fascicle => musculocutaneous nerve and median nerve

Branches of the lumbar plexus

Interested	Iliohypogastric nerve
In	Ilioinguinal nerve
Getting	Genitofemoral nerve
Laid	Lateral femoral cutaneous nerve
On	Obturator nerve
Fridays?	Femoral nerve

Anal sphincter innervation

L2, 3, 4 keep shit off the floor	L2, 3, 4

Diaphragm innervation

3, 4, 5 keeps the diaphragm alive	C3, C4, C5

Nerves passing through the jugular foramen

Vag you!	V – vagus, a – accessorius, g – glossopharyngeus

Language centers of CNS

Wernicke's area	Temporal lobe, sensory
Broca's area	Frontal lobe, motor

Nerve failure and hand positions

Hand of benediction	Median nerve
Claw hand	Ulnar nerve
Wrist-drop	Radial nerve

Eye

Layers of eye

The eyeball has three layers:

Outer fibrous layer (dura)	Sclera and cornea
Middle vascular layer (pia)	Choroid, ciliary body, iris
Inner layer	Pigmented epithelium, detachment of retina, retina

Retina

1) Neuroepithelial layer
   (outer and inner segment of receptors)
2) External limiting membrane
3) External nuclear layer
   (nuclei of rods and cones)
   1. NEURON
4) External plexiform layer
   (with horizontal cells = Interneurons)
5) Internal nuclear layer
   (nuclei of bipolar neurons)
   2. NEURON and Müller supporting cells
6) Internal plexiform layer
   (with amacrine cells = Interneurons)
7) Ganglionic layer (optic nerve)
   3. NEURON
8) Neurofibrous layer (axons)
9) Inside, internal limiting membrane

Anatomical and Medical Terminology

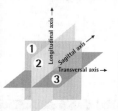

Planes of body	
1. Frontal plane	vertical planes passing through the body at right angles, dividing it into anterior and posterior
2. Sagittal plane	vertical planes passing through the body parallel to the median plane
3. Horizontal plane	planes passing through the body at right angles to the median and frontal planes, dividing it into superior and inferior

Commonly used terms of relationship

anterior	front of the body	lateral	farther from median plane	superior	nearer to head
posterior	back of the body	medial	nearer to median plane	deep	farther from surface
caudal	nearer to the coccyx area	dorsal	nearer to forefront of the body	superficial	nearer to or on surface
cranial	nearer to the head			palm	palmar surface of hand
distal	farther from trunk or point of origin (e.g., of a limb)	ventral	nearer to backside of the body	dorsum	dorsal surface of hand or foot
proximal	nearer to trunk or point of origin (e.g., of a limb)	inferior	nearer to feet		

Directional terms

ab-	away from	epi-	above, over, upon	per-	through, throughout
ad-	toward, near	eso-	within	peri-	around, surrounding
ambi-/ amphi-	around, on both sides, about	ex-	out, away from	post-	after, behind
ana-	up, backward, against	exo-	outside, outward	pre-	before, in front of
ante-	before, forward	extra-	outside	pro-	before
anter/o	front	hyper-	above, excessive, beyond	pros/o	forward, anterior
anti-	against	hypo-	under, deficient, below	proxim/o	near
apo-	away, separation	infra-	below, beneath	re-	back, again
circum-	around	inter-	between	retro-	behind, backward
contra-	against, opposite	intra-	within	sinistr/o	left
dextr/o	right	juxta-	near	sub-	under, beneath
dia-	through, throughout	later/o	side	super-	above, beyond
dis-	apart, to separate	levo-	left	supra-	above, beyond
dist/o	distant	medi/o	middle	tel/e	distant
ec-, ecto-	outside, out	meso-	middle	trans-	across
en-, endo-	inside, within	para-	alongside, near, beyond, abnormal	ultra-	beyond, excess

Numerical values

demi-, hemi-, semi-	half	mega-	M	million (10^6)
mono-, uni-	one	giga-	G	billion (10^9)
bi-, di-	two	tera-	T	trillion (10^{12})
tri-	three	peta-	P	quadrillion (10^{15})
tetra-, quadri-	four	exa-	E	quintillion (10^{18})
pent-, penta-, quinque-	five	deci-	d	one tenth (10^{-1})
hex-, hexa-, sex-	six	centi-	c	one hundreth (10^{-2})
hepta-, sept-, septi-	seven	milli-	m	one thousandth (10^{-3})
octa-, octi-	eight	micro-	µ	one millionth (10^{-6})
noni-	nine	nano-	n	one billionth (10^{-9})
deca-	da	ten (10^1)	pico-	p	one trillionth (10^{-12})
hekto-	h	hundred (10^2)	femto-	f	one quadrillionth (10^{-15})
kilo-	k	thousand (10^3)	atto-	a	one quintillionth (10^{-18})

Surgical procedures

-centesis	surgical punction	-plasty	surgical correction/repair	-tomy	surgical incision
-desis	surgical fixation, fusion	-rrhaphy	suture	-tripsy	to crush, break
-ectomy	surgical removal	-sect	to cut		
-pexy	fixation	-stomy	surgical opening		

Diagnostic procedures

aspir/o, aspirat/o	removal	-graphy	process of recording
-assay	to examine, to analyze	-meter	instrument for measuring
auscult/o, auscultat/o	to listen	-metry	process of measuring
echo-	reverberating sound	-opsy	to view
electr/o	electricity	palp/o, palpat/o	to touch gently
-gram	written record	percuss/o	to tap
-graph	instrument for recording	radi/o	x-ray, radiation

Pathogens

acar/o	mites	helminth/o	worm	myc/o	fungus	scolec/o	worm
bacteri/o-	bacteria	hirud/i	leech	parasit/o	parasite	vir/o	virus
fung/i	fungus, mushroom	ixod/i	ticks	pedicul/o	louse		

Colors

albin/o	white	erythr/o	red	lute/o	yellow	rubr/o	red
chlor/o	green	flav/o	yellow	melan/o	black	tephr/o	gray
cirrh/o	orange-yellow	fusc/o	dark brown	poli/o	gray	xanth/o	yellow
cyan/o	blue	glauc/o	gray	rhod/o	red		
eosin/o	red, rosy	leuk/o	white	rose/o	rosy		

The five senses

-opia	Vision	acous/o	Hearing
-geusia	Taste	olfact/o	Smell
haph/e	Touch		

Glossary

abdomin/o	abdomen
ablat/o	to remove
abras/o	to scrape off
actin/o	ray, radiation
-acut/o	sharp, severe
aden/o	gland
adip/o	fat
adren/o	adrenal glands
aer/o	air
agglutin/o	clumping
agit/o	rapidity, restlessness
-algesia	pain sensitivity
-algia	pain
allo-	other, different
alveol/o	alveolus
ambly/o	dim, dull
an/o	anus
andr/o	male
angi/o	vessel
anis/o	unequal
ankyl/o	stiff, crooked
antr/o	antrum
aphth/o	ulcer
apic/o	apex
arch/i, arch/e	first
arteri/o	artery
arteriol/o	arteriole
aspirat/o	inhaling, removal
asthenia	weakness
ather/o	fatty substance
atresia	closure
atrio	atrium
auto-	self
axill/o	armpit
azot/o	nitrogen, urea
balan/o	glans penis
bar/o	weight, pressure
bas/o	base, foundation
bil/i	bile
bio-	living
blast/o	immature, early stage of embryo
blenn/o	mucus
blephar/o	eyelid
brachi/o	arm
brachy/o	short
brady-	slow
bucc/o	cheek
burs/o	bursa
cac/o	bad, ill
calcane/o	heel
calor/i	heat
cardi/o	heart
carcin/o	cancer
cata-	down, under
cavit/o	hollow, cavity
cephal/o	head
cerebell/o	cerebellum
cerebr/o	cerebrum, brain
cervic/o	head, neck
-chalasia	relaxation
cheil/o	lip
chol/o	bile
chondr/o	cartilage
chrom/o	color
chron/o	time
cine-	movement
-clasis, -clasia	break
cleid/o	clavicle
coccyg/o	coccyx
cochle/o	cochlea
col/o	colon
colp/o	vagina
com-, con-	with, together
conjunktiv/o	conjunctiva
contus/o	to bruise
cor/o	pupil
corne/o	cornea
coron/o	heart
corpor/o	body
cost/o	rib
cox/o	hip
crani/o	skull
crin/o	secrete
cubit/o	elbow
cutane/o	skin
cry/o	cold
dacry/o	tear
dactyl/o	digit
derm/o, dermat/o	skin
desm/o	ligament
didym/o	a twin
dilat/o	expand
dipl/o	double
dips/o	thirst
dolor/o	pain
drom/o, -drome	running
duct/o	to lead
duoden/o	duodenum
-dynia	pain
dys-	bad, painful
-edema	swelling
-ectasis, -ectasia	expansion
-emesis	vomiting
encephal/o	brain
enter/o	(small) intestines
epididym/o	epididymis
epiglott/o	epiglottis
episi/o	vulva
erg/o	work
eu-	good, normal
faci/o	face
febr/i	fever
fet/o	fetus
fibr/o	fiber
fibul/o	fibula
fil/o, fil/i, filament/o	thread
flex/o, flect/o	bend
flu/o, flux/o	to flow
-form	specified shape
frig/o, frigid/o	cold
funct/o	performance
galact/o	milk
gamet/o	gamete
gangli/o, ganglion/o	ganglion
gastr/o	stomach
-gen, gen/o	generating
genit/o	reproduction
ger/o, geront/o	aged
gest/o, gestat/o	to bear
gingiv/o	gums
gloss/o	tongue
gluc/o, glyc/o	glucose
gnath/o	jaw
gnos/o	knowledge
gonad/o	gonads
-grade	step
gravid/o	pregnancy
gynec/o	female
hepat/o	liver
heredo-	heredity
hetero-	different
hidr/o	sweat

hist/o	tissue	narc/o	numbness	phag/o, -phagia	eating, ingestion
holo-	entire	nas/o	nose	phalang/o	phalanges
homeo-	likeness	nat/o	birth	pharmac/o	drugs
homo-	same, similar	necr/o	death	phas/o, -phasia	speech
humer/o	humerus	neo-	new	phen/o	appearance
hyal/o	resembling glass	nephr/o	kidney	-phil, -philia	affinity for
hydr/o	water	neur/o	nerve	phim/o	muzzle
hymen/o	hymen	noci-	to cause pain	phleb/o	vein
hypn/o	sleep	nod/o	knot	-phobia, phob/o	fear
hyster/o	uterus	norm/o	custom	phon/o, -phonia	voice, sound
iatr/o	treatment, physician	nos/o	disease	-phore, phor/o	bearer
ile/o	ileum	nutri/o, nutrit/o	nourish	phot/o	light
ir/o, irid/o	iris	ocul/o	eye	phren/o	mind, diaphragm
ischi/o	ischium	odont/o	tooth	phyl/o	species
is/o	equal	oesophag-	oesophagus	-phylaxis	protection
-itis	inflammation	-oid	resembling	physic/o	physical, normal
jejun/o	jejunum	olig/o	few, little	-physis	growth, growing
kyph/o	humpback	-oma	tumor	phyt/o, -phyte	plant
lacrim/o	tear	om/o	shoulder	pil/o	hair
lact/o	milk	omphal/o	navel	pin/o	to drink
lamin/o	lamina	oophor/o	ovary	plagi/o	oblique
lapar/o	abdomen	ophthalm/o	eye	plan/o	flat
laxat/o	to slacken	opisth/o	backward, behind	-plasty	surgical correction
lei/o	smooth	or/o	mouth	-plegia	paralysis
-lemma	confining membrane	orch/o, orchi/o, orchid/o	testis	pluri-	more, several
lept/o	thin, delicate	-orexia	appetite	pneum/o, pneumon/o	lung, air
lingu/o	tongue	orth/o	straight, normal	pod/o	foot
lip/o	fat	-osis	condition	-poiesis	formation
lith/o	stone	oste/o	bone	poikil/o	irregular
lob/o	lobe	pachy-	thick	poly-	many, much
log/o, -log, logue	speech	ped/o	foot, child	-porosis	decrease in density
-logy	study of	palat/o	palate	potenti/o	power, strenght
lord/o	curvature	palin-, pali-	recurrence	-prandial	meal
lumin/o	light	palpebr/o	eyelid	primi-, prot/o	first
luxat/o	dislocate	pan-	all	proct/o	rectum, anus
-lysis	dissolution	pancreat/o	pancreas	prosop/o	face
mamm/o, mast/o	breast	papul/o	papule, pimple	prostat/o	prostate gland
-malacia	sofeting	-para, -parous	to bear	pseudo-	false
maxill/o	maxilla	parathyroid/o	parathyroid	-ptosis	prolapse, drooping
mediastin/o	mediastinum	-paresis	partial paralysis	ptyal/o	saliva
medull/o	marrow	-partum	childbirth	pub/o	pubis
mega-, megalo-	large	patell/o	patella	pulmon/o	lung
mening/o	meninges	path/o	disease	puls/o, pulsat/o	to beat, beating
mer/o	part	pector/o	chest	purul/o	pus formation
morph/o	shape	pelv/i	pelvis	pykn/o, pycn/o	thick, dense
mort/o	death	-penia	deficiency	py/o	pus
-motor	movement	perine/o	perineum	pyret/o	fever
muc/o, myx/o	mucus	peritone/o	peritoneum	pyr/o	heat
my/o, myos/o	muscle	-petal	moving toward	ram/i	branch

-(re)ceptor	receiver
rect/o	rectum
relaps-	to slide back
respir/o, respirat/o	breathe, breathing
retin/o	retina
retract/o	drawing back
rhiz/o	root
rot/o, rotat/o	turn
-rrhagia, -rrhage	profuse fluid discharge
-rrhexis	rupture
-rhe/o	flow, current, stream
sapr/o	rotten, decay
sarc/o	flesh
scapul/o	scapula
scat/o	feces
schist/o, -schisis, schiz/o	split
scler/o	sclera
scoli/o	crooked, twisted
seb/o	sebum
sedat/o	to calm
semin/i	semen
senil/o	old (age)
sensor/i	sensory
sens/o, sensat/o	feeling, perception
-sepsis, septic/o	putrefaction
sept/o	partition
ser/o	serum
sial/o	saliva
sicc/o	to dry
sinus/o, sin/o	cavity, sinus
soci/o	social, society
solut/o	dissolved
somat/o	body
somn/i, -somnia	sleep
son/o	sound
-spasm, spasm/o	involuntary contraction
spectr/o	image, spectrum
spher/o	round, sphere
sphygm/o, -sphyxia	pulse
spin/o	spinal cord
spir/o	breathe
splanchn/o	viscera
splen/o	spleen
spongi/o	spongelike
soci/o	social, society
spor/o	spore, seed
squam/o	squamous, scales
-stasis	standing still
steat/o	fat
-stenosis, sten/o	narrowed
stere/o	solid
steril/o	barren
sthen/o, -stenia	strength
strat/o	layer
strict/o	to tighten
-stroma	supporting tissue of an organ
sulc/o	furrow, groove
sym-, syn-	with, together
synaps/o, synapt/o	point of contact
tachy-	fast
tact/o	touch
tal/o	talus
tars/o	tarsus, edge of eyelid
-taxia, tax/o	arrangement, coordination
tect/o	rooflike
tegment/o	covering
tel/o	end
temp/o, tempor/o	period of tome
ten/o, tenont/o	tendon
-tension, tens/o	stretched, strained
terat/o	monster
termin/o	boundary, limit
thanat/o	death
thec/o	sheath
-therapy, therapeut/o	treatment
therm/o	heat
thorac/o	chest
thromb/o	clot, thrombus
-thymia	mind, emotions
thyr/o	thyroid gland
tibi/o	tibia
ton/o	tone, tension
tonsill/o	tonsils
top/o	particular place or area
tors/o	twisting, twisted
tox/o, toxic/o	poison
trache/o	trachea
trachy-	rough
traumat/o	trauma, injury, wound
trem/o, tremul/o	shaking
-tresia	opening, perforation
trich/o	hair
-trophy	nourishement
troph/o	growth
tub/o	tube
tympan/o	eardrum
-type, typ/o	representative form
typhl/o	cecum, bindness
un-	not, reversal
ungu/o	nail
-uresis	urination
-uria	urine condition
ur/o, urin/o	urine
uter/o	uterus
uve/o	uvea
vag/o	vagus nerve
-valv(ul)/o	valve
-vari(at)/o	change, vary
vas/o	vessel, vas defens
vascul/o	blood vessel
ventil/o	to aerate
ventricul/o	ventricle of the heart or brain
verruc/i	wart
vers/o, -verse	turn, turning
vertebr/o	vertebra
vestibul/o	vestibule
viscer/o	internal organs
viv/i	life, alive
-volemia	blood volume
volv/o, volut/o	to roll
vulv/o	vulva
xen/o	strange, foreign matter
xer/o	dry
zon/i, zon/o	zone
zo/o	animal
zyg/o	union, junction
zym/o	enzyme, ferment

W

Z

Notes

Notes

Börm Bruckmeier Products

pockets

Acupuncture pocket	ISBN 978-1-59103-248-9	US $ 16.95
Anatomy pocket	ISBN 978-1-59103-219-9	US $ 16.95
Canadian Drug pocket 2009	ISBN 978-1-59103-238-0	US $ 14.95
Differential Diagnosis pocket	ISBN 978-1-59103-216-8	US $ 14.95
Drug pocket 2008	ISBN 978-1-59103-240-3	US $ 12.95
Drug pocket plus 2008	ISBN 978-1-59103-241-0	US $ 19.95
ECG pocket	ISBN 978-1-59103-230-4	US $ 16.95
ECG Cases pocket	ISBN 978-1-59103-229-8	US $ 16.95
Homeopathy pocket	ISBN 978-1-59103-250-2	US $ 14.95
Medical Abbreviations pocket	ISBN 978-1-59103-221-2	US $ 16.95
Medical Classifications pocket	ISBN 978-1-59103-223-6	US $ 16.95
Medical Spanish pocket	ISBN 978-1-59103-232-8	US $ 16.95
Medical Spanish Dictionary pocket	ISBN 978-1-59103-231-1	US $ 16.95
Medical Spanish pocket plus	ISBN 978-1-59103-239-7	US $ 22.95
Medical Translator pocket	ISBN 978-1-59103-235-9	US $ 16.95
Normal Values pocket	ISBN 978-1-59103-205-2	US $ 12.95
Nursing Dictionary pocket	ISBN 978-1-59103-237-3	US $ 12.95
Respiratory pocket	ISBN 978-1-59103-228-1	US $ 16.95
Wards 101 pocket	ISBN 978-1-59103-253-3	US $ 19.95

pocketcards

Acute Coronary Syndrome	ISBN 978-1-59103-073-7	US $ 3.95
Alcohol Withdrawal pocketcard	ISBN 978-1-59103-031-7	US $ 3.95
Anesthesiology pocketcard Set (3)	ISBN 978-1-59703-050-8	US $ 9.95
Antibiotics pocketcard 2009	ISBN 978-1-59103-064-5	US $ 3.95
Antibiotics pocketcard Set (2)	ISBN 978-1-59103-056-0	US $ 6.95
Antifungals pocketcard	ISBN 978-1-59103-013-3	US $ 3.95
Antithrombotic Therapy pocketcard Set (2)	ISBN 978-1-59103-070-6	US $ 6.95
Asthma pocketcard Set (2)	ISBN 978-1-59103-046-1	US $ 6.95
COPD pocketcard Set (2)	ISBN 978-1-59103-047-8	US $ 6.95
Dementia pocketcard Set (3)	ISBN 978-1-59103-053-9	US $ 9.95
Diabetes pocketcard Set (3)	ISBN 978-1-59103-054-6	US $ 9.95
Depression pocketcard Set (3)	ISBN 978-1-59103-067-6	US $ 9.95
Dyslipidemia pocketcard Set (2)	ISBN 978-1-59103-055-3	US $ 6.95
ECG pocketcard	ISBN 978-1-59103-028-7	US $ 3.95
ECG Ruler pocketcard	ISBN 978-1-59103-002-7	US $ 3.95
ECG pocketcard Set (3)	ISBN 978-1-59103-003-4	US $ 9.95
Echocardiography pocketcard Set (2)	ISBN 978-1-59103-024-9	US $ 6.95
Epilepsy pocketcard Set (2)	ISBN 978-1-59103-034-8	US $ 6.95
Geriatrics pocketcard Set (3)	ISBN 978-1-59103-037-9	US $ 9.95
Heart Failure pocketcard Set (2)	ISBN 978-1-59103-071-3	US $ 6.95
History & Physical Exam pocketcard	ISBN 978-1-59103-022-5	US $ 3.95
Hypertension pocketcard	ISBN 978-1-59103-042-3	US $ 3.95

Order: www.media4u.com ◆ Call: 1-800-266-5564 www.media4u.com

Börm Bruckmeier Products

pocketcards

Medical Abbreviations pc Set (2)	ISBN 978-1-59103-010-2	US $ 6.95
Medical Spanish pocketcard	ISBN 978-1-59103-027-0	US $ 3.95
Medical Spanish pocketcard Set (2)	ISBN 978-1-59103-025-6	US $ 6.95
Metabolic Syndrome pocketcard Set (3)	ISBN 978-1-59103-065-2	US $ 9.95
Migraine pocketcard Set (3)	ISBN 978-1-59103-066-9	US $ 9.95
Nephrology pocketcard Set (3)	ISBN 978-1-59103-061-4	US $ 9.95
Neurology pocketcard (3)	ISBN 978-1-59103-045-4	US $ 9.95
Normal Values pocketcard	ISBN 978-1-59103-023-2	US $ 3.95
Obesity pocketcard Set (3)	ISBN 978-1-59103-075-1	US $ 9.95
Oncology pocketcard Set (3)	ISBN 978-1-59103-059-1	US $ 9.95
Osteoporosis pocketcard Set (2)	ISBN 978-1-59103-072-0	US $ 6.95
Pain Management pocketcard Set (2)	ISBN 978-1-59103-076-8	US $ 6.95
Parkinson pocketcard Set (2)	ISBN 978-1-59103-043-0	US $ 6.95
Periodic Table pocketcard	ISBN 978-1-59103-014-0	US $ 3.95
Psychiatry pocketcard Set (2)	ISBN 978-1-59103-033-1	US $ 6.95
Pulmonary Function Test pocketcard Set (2)	ISBN 978-1-59103-074-4	US $ 6.95
Rheumatoid Arthritis pocketcard Set (2)	ISBN 978-1-59103-048-5	US $ 6.95
Urinary Incontinence pocketcard	ISBN 978-1-59103-069-0	US $ 3.95
Vision pocketcard	ISBN 978-1-59103-032-4	US $ 3.95
Wound Ruler pocketcard	ISBN 978-1-59103-051-5	US $ 3.95

pockettools

Asthma pockettool	ISBN 978-1-59103-802-3	US $ 9.95
COPD|PFT pockettool	ISBN 978-1-59103-806-1	US $ 9.95
DARF pockettool	ISBN 978-1-59103-803-0	US $ 9.95
ECG pockettool	ISBN 978-1-59103-800-9	US $ 9.95
ECG Ruler pockettool	ISBN 978-1-59103-805-4	US $ 9.95
Medical Spanish pockettool	ISBN 978-1-59103-804-7	US $ 9.95
Normal Values pockettool	ISBN 978-1-59103-801-6	US $ 9.95

pocketflyer

Diabetes mellitus pocketflyer	ISBN 978-1-59103-850-4	US $ 9.95

PDA software

Differential Diagnosis pocket for PDA	ISBN 978-1-59103-600-5	US $ 16.95
Drug Therapy pocket for PDA	ISBN 978-1-59103-605-0	US $ 16.95
ECG pocket for PDA	ISBN 978-1-59103-601-2	US $ 16.95
Homeopathy pocket for PDA	ISBN 978-1-59103-650-0	US $ 16.95
ICD-9-CM 2005 for PDA	ISBN 978-1-59103-606-7	US $ 24.95
Medical Abbreviations pocket for PDA	ISBN 978-1-59103-603-6	US $ 16.95
Medical Calculator pocket for PDA	ISBN 978-1-59103-616-6	US $ 16.95
Medical Spanish pocket for PDA	ISBN 978-1-59103-602-9	US $ 16.95
Medical Spanish Dic. pocket for PDA	ISBN 978-1-59103-607-4	US $ 16.95
Medical Spanish pocket plus for PDA	ISBN 978-1-59103-608-1	US $ 24.95